National Climate Change Adaptation

EMERGING PRACTICES IN MONITORING AND EVALUATION

This work is published on the responsibility of the Secretary-General of the OECD. The opinions expressed and arguments employed herein do not necessarily reflect the official views of the Organisation or of the governments of its member countries.

This document and any map included herein are without prejudice to the status of or sovereignty over any territory, to the delimitation of international frontiers and boundaries and to the name of any territory, city or area.

Please cite this publication as:
OECD (2015), *National Climate Change Adaptation: Emerging Practices in Monitoring and Evaluation*, OECD Publishing, Paris.
http://dx.doi.org/10.1787/9789264229679-en

ISBN 978-92-64-22966-2 (print)
ISBN 978-92-64-22967-9 (PDF)

The statistical data for Israel are supplied by and under the responsibility of the relevant Israeli authorities. The use of such data by the OECD is without prejudice to the status of the Golan Heights, East Jerusalem and Israeli settlements in the West Bank under the terms of international law.

Photo credits: © iStockphoto.com/Laures.

Corrigenda to OECD publications may be found on line at: *www.oecd.org/publishing/corrigenda*.

Foreword

As the international community prepares to negotiate a new climate deal in Paris in December 2015, the consequences of growing concentrations of greenhouse gases are becoming increasingly apparent. The Earth's surface temperature has been successively warmer over the last three decades than any decade since 1850. This is contributing to changes in precipitation patterns as well as sea level rise and increases in the frequency and intensity of temperature extremes.

The international community has recognised the urgency of building resilience against the effects of climate variability and change. The OECD has been supporting this process by developing guidance on adaptation planning in both developed and developing countries. This report, National Climate Change Adaptation: Emerging Practices from Monitoring and Evaluation, *proposes a number of practical tools that governments may draw upon for this purpose.*

Adapting to a changing climate, involves decision making "with continuing uncertainty about the severity and timing of climate change impacts", according to the latest analysis by the Intergovernmental Panel on Climate Change. In this uncertain environment, a flexible approach to adaptation planning and implementation can benefit from continuous learning from monitoring and evaluation. Furthermore, the information generated from monitoring and evaluation can inform national approaches to adaptation that are robust and applicable to a range of possible future climate outcomes.

With continuing constraints on government budgets, it is also vital to ensure that interventions to build resilience to climate change are well targeted and deliver agreed objectives. While this is in countries' own interest, it will also demonstrate at the international level that resources allocated for adaptation are effective in reducing vulnerability to the effects of climate change at the local and national level. The tools proposed in this report can help governments identify which approaches to adaptation are effective in achieving agreed objectives and to shed light on some of the enabling factors for their success.

Promoting climate resilient development is only possible by learning what approaches to adaptation are effective and using that knowledge in domestic planning and budgeting processes. The OECD stands ready to support countries in their efforts to put in place effective national adaptation plans and the related monitoring and evaluation frameworks.

Angel Gurría
OECD Secretary-General

Acknowledgements

This report *National Climate Change Adaptation: Emerging Practices in Monitoring and Evaluation* is an output from the OECD Task Team on Climate Change and Development Co-operation that is overseen jointly by the Working Party on Climate, Investment and Development (WPCID) of the Environment Policy Committee (EPOC) and the Network on Environment and Development Co-operation (ENVIRONET) of the Development Assistance Committee (DAC).

This report has been written by Nicolina Lamhauge under the supervision of Michael Mullan. Anthony Cox provided oversight and valuable feedback. In addition to members of the Task Team on Climate Change and Development Co-operation, the author would like to thank Joëlline Benefice, Simon Buckle, Juan Casado-Asensio, Jan Corfee-Morlot, Jane Ellis, Eva Hübner, Megan Grace Kennedy-Chouane, Britta Labuhn, Hans Lundgren, and Alexis Robert of the OECD for valuable input and feedback. Janine Treves and Katherine Kraig-Ernandes provided editorial support.

Financial contributions from the UK Department for International Development and the Swiss Development Co-operation Agency are gratefully acknowledged.

Table of contents

Acronyms ... 7
Executive summary ... 9

Part I

Ensuring effective adaptation to climate change

Chapter 1. Assessing national climate change adaptation ... 13
- Objectives of national monitoring and evaluation of adaptation ... 15
- Challenges to the monitoring and evaluation of adaptation ... 19
- Notes ... 25
- References ... 25

Chapter 2. Effective monitoring and evaluation of climate change adaptation ... 29
- Data availability and monitoring and evaluation capacity ... 30
- Co-ordination between providers and users of climate information ... 36
- Note ... 38
- References ... 38

Chapter 3. National tools for monitoring and evaluation of climate change adaptation ... 39
- Climate change risk and vulnerability assessments ... 40
- Indicators for monitoring progress against adaptation priorities ... 44
- Learning from adaptation approaches ... 49
- National audits and climate expenditure reviews ... 52
- Notes ... 58
- References ... 58

Part II

Emerging country indicators to monitor and evaluate adaptation

Chapter 4. Proposed indicators for Kenya's climate change action plan ... 63

Chapter 5. Goals and outcomes in Philippines' climate change action plan ... 71

Chapter 6. Indicators used to evaluate adaptation in the United Kingdom ... 81

Chapter 7. Proposed Indicators for monitoring the German adaptation strategy ... 85

Chapter 8. Australia's proposed climate adaptation assessment framework ... 89

Chapter 9. Measures and actions in France's national adaptation plan ... 93

Tables

1.1. Typology of approaches for adaptation monitoring and evaluation 22
1.2. Examples of national monitoring and evaluation frameworks for adaptation 24
2.1. Domestic sources of data to monitor and evaluate climate change adaptation in Kenya 33
2.2. Examples of capacity building elements for monitoring and evaluation 34
3.1. Prioritising climate change risks 41
3.2. Examples of data and information that can be used in risk and vulnerability assessments 42
3.3. Types of indicators used to monitor and evaluate adaptation 47
3.4. Questions and information requirements to meet indicator standards 48
3.5. Key questions for adaptation audits 54
3.6. Questions to consider for a Climate Public Expenditure and Investments Review 56

Acronyms

ASC	Adaptation Sub-Committee
CCRA	Climate Change Risk Assessment
CLE	Country-led evaluation
CPEIR	Climate Public Expenditure and Institutional Review
CRS	Creditor Reporting System
DAC	Development Assistance Committee
EUROSAI	European Organisation of Supreme Audit Institutions
ICF	International Climate Fund
INTOSAI	International Organisation for Supreme Audit Institutions
KCCAP	Kenya Climate Change Action Plan
LDC	Least developed country
MfDR	Managing for Development Results
NAP	National Adaptation Plan
NAPA	National Adaptation Programmes of Action
NCCAP	National Climate Change Action Plan
NPBMF	National Performance and Benefits Measurement Framework
PAF	Performance Assessment Framework
PAP	Programme Aid Partnership
PPCR	Pilot Program for Climate Resilience
SAI	Supreme audit institutions
TAMD	Tracking Adaptation and Measuring Development
UNFCCC	United Nations Framework Convention on Climate Change
WGEA	Working Group on Environmental Auditing

Executive summary

Countries' national approaches to climate change adaptation are increasingly moving from a project focus towards more integrated strategies that promote co-ordination across sectors and levels of government. The monitoring and evaluation frameworks assessing the effectiveness of the national approach on adaptation must be adjusted accordingly. With an integrated approach to adaptation, a country's resilience to climate change reflects the change brought about by individual adaptation interventions, as well as that caused by socio-economic trends and policies implemented for reasons other than climate change.

This report draws on the emerging practice of monitoring and evaluation of adaptation in developed and developing countries to identify four tools that can be used to enhance learning and to assess countries' progress in adapting to climate change. The report also considers the potential role development co-operation providers can play in helping partner countries to implement the four tools and build on the information they generate.

Learning and accountability are twin objectives of monitoring and evaluation of adaptation

Domestic efforts to adapt to climate change are at their strongest when they include a flexible process based on continuous learning from monitoring and evaluation. Further, monitoring and evaluation can help to ensure that resources earmarked for adaptation, or mainstreamed through other initiatives, contribute to agreed objectives in a cost-effective manner. The nature of this accountability mechanism, however, depends on countries' approaches to adaptation, the governance systems in place, and the financing mechanisms used. Theoretical frameworks have proposed how monitoring and evaluation can achieve the twin objectives of learning and accountability. In practice, national frameworks are constrained by domestic data availability and monitoring and evaluation capacity. Given the diverse set of initiatives contributing to a country's level of climate resilience, good co-ordination between the producers and the users of the information is important.

A portfolio of tools can contribute to a better understanding of changes in climate risks and resilience

A portfolio of monitoring and evaluation tools is needed to assess the impact of public and private, planned and autonomous adaptation initiatives. Separately, each tool will ideally capture a distinct component of the climate risks and vulnerabilities; combined they can contribute to a better overview of the larger picture. While the applicability of such tools will vary across countries and over time, the feasibility of applying them will also differ. The four tools examined in this report are not an exhaustive list, but instead represent promising avenues for further work based on countries' experiences to date:

- **Climate change risk and vulnerability assessments.** When conducted at the outset of a national focus on adaptation, such assessments can contribute to a baseline of the country's climate vulnerability against which progress on adaptation can be reviewed. If the assessments are repeated on a regular basis (e.g. to inform national planning and budgeting cycles) they can provide a picture of how climate risks and vulnerabilities are changing over time. However, to understand how these changes came about, the assessments can benefit from the application of complementary tools, including those outlined below.
- **Indicators to monitor progress on adaptation priorities.** Indicators can facilitate the monitoring of climate risks and vulnerabilities over time and between locations. Since the identification, collection, and use of indicators is resource intensive, a carefully defined set of qualitative and quantitative indicators may be aligned to the adaptation priorities identified in the country's strategic approach on adaptation. Alternatively, the indicator set may draw on existing datasets and, where possible, on indicators used to monitor and evaluate national development plans and policies. However, indicators alone will fail to provide adequate insight into, and understanding of, the context in which adaptation is taking place.
- **Project and programme evaluations to identify effective adaptation approaches.** Although the evaluations of adaptation projects and programmes face a number of challenges and uncertainties, they can help to identify what approaches to adaptation are effective in achieving agreed objectives. Further, they can contribute to a better understanding of the conditions required for the adaptation measures to succeed. Individual countries can benefit from lessons learned from large adaptation interventions and innovative pilot approaches to adaptation.
- **National audits and climate expenditure reviews.** These examine whether public expenditures on adaptation are aligned with national and international policy goals, are allocated in accordance with existing rules, regulations and principles of good governance, and if they are allocated in a cost-effective manner. Further, audits and expenditure reviews examine whether the national institutional mechanisms are in place to effectively manage and deliver climate finance. They support accountability, particularly in developing countries where resources received from development co-operation providers may be specifically earmarked for adaptation.

Development co-operation providers can support partner countries in the monitoring and evaluation of adaptation

Development co-operation providers can support partner countries in their efforts to monitor and evaluate adaption through, for example, peer reviews and the sharing of experiences between countries. To aid this process, development co-operation providers and partner countries must put in place systems that support monitoring and evaluation, and plan interventions in ways that readily facilitate learning throughout the process. At the same time, development co-operation providers can ensure that the data and information gathered for their own monitoring and evaluation is made publicly available. This can help reduce the risk of data collection measures being duplicated, especially in resource constrained countries.

PART I

Ensuring effective adaptation to climate change

National Climate Change Adaptation
Emerging Practices in Monitoring and Evaluation

PART I

Chapter 1

Assessing national climate change adaptation

This chapter examines the objectives and challenges of national monitoring and evaluation of climate change adaptation. It briefly reviews what such monitoring and evaluation frameworks may look like in theory and practice. The chapter also considers notions of climate risk, vulnerability and resilience, as well as the need to establish baselines and targets for monitoring and evaluation.

Key messages

- Continuous learning from monitoring and evaluation can help to inform the formulation of the national policy agenda on adaptation. This however, requires a flexible adaptation process that can respond to changing climate circumstances.
- The multifaceted nature of adaptation makes it essential to use a portfolio of monitoring and evaluation tools that generate lessons learned and can guide any mid-course adjustments that may be needed.
- Assessing the value for money of the resources allocated for adaptation is important, but it should not be the sole objective of monitoring and evaluation activities.
- To overcome challenges in monitoring and evaluating adaptation, countries may initially focus on progress made in addressing current climate vulnerability. As climate uncertainty decreases, and data availability and monitoring and evaluation capacity improve, the focus may gradually shift towards an evaluation of current levels of adaptation against projected climate change.
- Building on systems already in place to collect and process climate information can help to reduce administrative burdens and ensure sustainability.

National governments are increasingly taking action to support climate change adaptation. Fifty Least Developed Countries (LDCs) have formulated National Adaptation Programmes of Action (NAPAs), which identify the countries' urgent and immediate adaptation needs. These are now at varying stages of implementation (UNFCCC, n.a.). The National Adaptation Plan (NAP) process established under the United Nations Framework Convention on Climate Change (UNFCCC) in 2010 will continue to support LDCs and other developing countries in formulating their medium and long-term adaptation needs, bringing in a more strategic, national approach to complement the use of stand-alone projects and programmes (UNFCCC, 2011). Similarly, there has been an increase in adaptation planning in developed countries. Since the first OECD country published its national adaptation strategy in 2005, more than two thirds of the 34 OECD member countries now have national adaptation policies in place (Mullan et al., 2013). The most common approach to adaptation in developed countries has been to integrate it into all planning and budgeting processes, aiming to align adaptation duties with existing ministerial responsibilities. The NAP process intends to facilitate a similar approach in developing countries.

Most developing countries rely, at least in part, on external support to meet their adaptation needs. OECD countries play an important role as providers of such financial support. Bilateral financial commitments for adaptation-related interventions by members of the OECD Development Assistance Committee (DAC) averaged USD 9.3 billion per year between 2010 and 2012 (OECD, 2014). This support illustrates the mainstreamed nature of many adaptation interventions with general development objectives. For example, of total adaptation-related aid commitments, over 70% was mainstreamed into activities primarily motivated by development objectives other than adaptation.[1] This bilateral funding is complemented with multilateral financing and resources from dedicated green and climate funds.[2]

Despite progress made in defining, implementing and financing national adaptation priorities, the formulation of complementary monitoring and evaluation frameworks has generally lagged behind. Monitoring and evaluation are two separate but closely linked processes. Monitoring examines, on an on-going basis, progress made in implementing planned initiatives that directly or indirectly affect the level of climate resilience. Further, monitoring may, for example, entail a continuous assessment of the enabling environment in place for adaptation, and of the capacities to develop and implement adaptation policies, plans and strategies. Evaluation, on the other hand, is an independent assessment of progress made in reducing climate risks and vulnerabilities and an analysis of how the change came about. Evaluations are based on the data monitored, but they also draw on other relevant information such as stakeholder consultations and expert reviews. Lessons learned from monitoring and evaluation can guide any mid-course adjustments that may be needed of policies in place and inform subsequent measures. Monitoring and evaluation also ensure transparency around the allocation, use and results achieved through development support.

Box 1.1. **Climate risk, vulnerability and resilience**

This report uses the definition of adaptation to climate change proposed by the Intergovernmental Panel on Climate Change. It defines adaptation as the outcome of reduced exposure and vulnerability to climate risk and increasing resilience to the potential adverse impacts of climate extremes. The objective of a monitoring and evaluation framework for adaptation is therefore to assess if countries over time are able to reduce the exposure and vulnerability of people and infrastructure to natural climate variability and anthropogenic climate change.

Source: IPCC (2012), *Managing the Risks of Extreme Events and Disasters to Advance Climate Change Adaptation: A Special Report of Working Groups I and II of the Intergovernmental Panel on Climate Change*, Field et al. (eds.), Cambridge University Press, Cambridge and New York.

National monitoring and evaluation systems often try to achieve multiple objectives, yet the most suitable approach will depend on the particular context. This chapter explores the main objectives and challenges for monitoring and evaluating adaptation at the national level. Some of the challenges discussed equally apply to adaptation projects and programmes. The chapter also briefly reviews some theoretical approaches to monitoring and evaluating adaptation and compares these with the approaches currently being designed and implemented by countries.

Objectives of national monitoring and evaluation of adaptation

The objectives of monitoring and evaluating adaptation vary by country, but two common themes include learning and accountability. Learning aims to enhance stakeholders' understanding of the country's climate change risks and vulnerabilities that in turn can help to identify approaches that are effective in reducing those risks. Accountability aims to ensure that resources allocated for adaptation are effective in achieving set objectives.

Monitoring and evaluation for learning

Gaps remain in current understanding of climate change impacts and vulnerability. For example, the quality and usability of climate projections is uneven due to resource constraints and data limitations (OECD, 2009). As a result, average changes over wide areas can be relatively well understood, while there is greater uncertainty about the specific impacts at the local level, particularly in countries with diverse ecosystems or topography (e.g. Nepal and Mozambique). Regional climate models and statistical techniques have been developed that can downscale climate projections to provide a higher resolution (Ranger, Muir-Wood and Priya, 2009). However, applying such techniques requires technical capacity not available in all developing countries. Given the uncertainties inherent in climate projection, national adaptation policies and planning processes can benefit from periodic reviews and assessments.

Similarly, the effectiveness of adaptation measures is often poorly understood. By building on improved climate projections and lessons learned (from initiatives focusing specifically on adaptation, as well as those focusing on climate variability and disaster risk reduction), the capacity to adapt to climate change can gradually improve (IEG, 2013). The value of monitoring and evaluation of adaptation as a mechanism for learning therefore lies in the use of the information for adaptation planning processes and for improving government performance.

Despite the importance of monitoring and evaluation for learning, there are two significant barriers to achieving this goal (with a more exhaustive list summarised in Box 1.2). At the national level, it can be difficult to ensure that the lessons learned are readily available to, and used by, the stakeholders shaping the domestic policy agenda on adaptation (GEF IEO, 2013; Kato et al., 2014; OECD, 2001). In developing countries, an additional barrier to learning can be that the monitoring and evaluation systems to varying degrees may be shaped by the information required by the providers of climate finance rather than by the national authorities. This can create a disincentive for exploring opportunities for learning beyond those included in the initial funding agreement.

Box 1.2. **Barriers to learning from monitoring and evaluation**

There are a number of barriers to learning from monitoring and evaluation that apply equally to the national, project and programme levels:

- **Organisational culture:** In some organisational structures poor performance is associated with blame, discouraging openness and learning. Other structures see failure to deliver expected results as an opportunity for learning.
- **Pressure to spend:** Pressure to meet disbursement targets reduces the time available to examine lessons learned and to integrate them in the planning process.
- **Lack of incentives to learn:** When staff turn-over is high, the incentive to learn may be limited since the staff responsible will often have moved on long before the consequences of failure to learn are felt.
- **Tunnel vision:** Some staff or operational units prefer to stick to their old processes and procedures even when the shortcomings of these approaches are recognised.
- **Loss of institutional memory:** The organisational capacity to use monitoring and evaluation as a mechanism for learning may be reduced when staff turn-over is high.

Box 1.2. **Barriers to learning from monitoring and evaluation** *(cont.)*

- **Insecurity and the pace of change:** Unclear and frequent shifts in priorities can have an adverse effect on learning.
- **Unequal nature of relationship:** The unequal relationship between development co-operation providers and partner countries can inhibit two-way knowledge sharing.

Source: OECD (2001), *Evaluation Feedback for Effective Learning and Accountability*, OECD Publishing, Paris.

Political commitment and buy-in is important for overcoming these barriers. Ministers and other senior policy officials can champion the importance of monitoring and evaluation and ensure that findings contribute to a transparent, evidence-based adaptation policy planning and implementation process (Segone, 2008). There is also scope for greater exchange among countries of lessons learned on effective adaptation approaches and on methods used to monitor and evaluate them.

Monitoring and evaluation as an accountability measure

There are two dimensions to the use of monitoring and evaluation as an accountability measure: answerability and enforceability. Answerability is primarily based on political will to justify decisions and actions based on the monitoring and evaluation of adaptation interventions. Enforceability, on the other hand, refers to the ability of governments to ensure that national policy commitments agreed upon (e.g. in their annual or periodic development plans) are being met and that corrective measures are undertaken when they are not. There are three broad categories of enforceability (SADEV, 2012):

- **Representative enforceability** between elected representatives and citizens managed through the democratic process of elections, free access to information and legislative oversight of the executive;
- **Corporate enforceability** through a legally binding contract (e.g. between development co-operation providers and partner countries), where the primary emphasis is on compliance of contract agreements;
- **Collaborative enforceability** that is not based on a political or legal commitment but rather on shared interests and commitments to achieve a common goal.

In addition to domestic answerability and representative enforceability, developing countries that receive support from development co-operation providers may also face corporate enforceability. This means that access to support is based in part on the countries' ability to demonstrate that resources are effectively allocated and that agreed objectives are being achieved (SADEV, 2012; UNDP, 2010). Some developing countries may face such corporate enforceability from multiple providers of co-operation, limiting the resources available to establish domestic monitoring and evaluation systems that focus on domestic learning and accountability needs.

Mutual accountability can help to ensure that developing countries remain primarily accountable to their own citizens. This concept was introduced in the 2005 *Paris Declaration on Aid Effectiveness* and reiterated in the 2008 *Accra Agenda for Action* and the 2011 *Busan Partnership for Effective Development Co-operation*. The objective of mutual accountability is to facilitate a process whereby development co-operation providers and partner countries are held jointly accountable to a set of agreed commitments (OECD, n.a.). In practice, achieving this can be challenging given the domestic oversight development co-operation providers face. For example, an independent evaluation of the implementation of the *Paris Declaration* found that results management and mutual accountability were two of the areas with the least progress made by development partners (Wood et al., 2011). Similarly, a review of the Global Partnership for Effective Development Co-operation found that targeted efforts are needed to make mutual review processes more transparent and inclusive, extending participation to emerging providers, civil society organisations and the private sector (OECD/UNDP, 2014). Box 1.3 summarises an example of mutual accountability in Mozambique and the possible limitations.

Box 1.3. **Programme Aid Partners Performance Assessment Framework**

In Mozambique, a Performance Assessment Framework (PAF) has been jointly agreed upon by the Government of Mozambique and 19 bilateral and multilateral development agencies that are signatories to the Programme Aid Partnership (PAP). Through the PAF, PAP members identify 35 socio-economic targets to be achieved within a set period of time, usually three to four years. The targets are based on national development objectives identified e.g. in the national poverty reduction strategy and Mozambique's Five-Year Programme. PAP members jointly assess performance in achieving agreed objectives, and development co-operation providers are assessed on their performance in meeting the aid effectiveness principles. Discussion is currently underway to include a strategic objective on climate change measured against the following indicator: "Cumulative number of sectors/institutions and provinces that integrate disaster risk management, climate change adaptation and mitigation aspects into planning processes." The indicator will include annual targets for sectors and provinces.

The PAF is intended to reduce the need for different reporting requirements. In practice, however, most development partners do not provide all their assistance through budget support but also finance stand-alone projects and programmes that usually have their own reporting requirements. Further, members of the PAP account for just over a third of all development support to Mozambique. Some of the current members (e.g. Belgium, the Netherlands and Spain) have announced that they will end their budget support, while others (e.g. Sweden) are becoming more reluctant to provide this form of support. Lastly, some members of the PAP (e.g. the UN and USAID) do not provide budget support but rather support projects and loans. A possible consequence of these trends is that Mozambique in the future will face an increase in the number of project or programme specific monitoring and evaluation requirements. While these may contribute to gradual learning and enhanced monitoring and evaluation capacity, they may also deter domestic resources away from strengthening the national approach to monitoring and evaluation.

Source: Handley, G. (2008), *Mutual Accountability at the Country Level: Mozambique Case Study*, ODI, London; SADEV (2012), *Mutual accountability in practice: The case of Mozambique*, Swedish Agency for Development Evaluation, Karlstad; IIED (2013b), *Tracking Adaptation and Measuring Development (TAMD) in Mozambique: Appraisal and Design Phase Report*, International Institute for Environment and Development, London.

Some developing countries report on separate results frameworks that meet the individual reporting requirements of their development co-operation providers (IIED, 2013a; GIZ, 2013). While this may, in the long-term, enhance the country's domestic monitoring and evaluation capacity, it is resource intensive and can divert attention and domestic resources from ensuring answerability and representative enforceability. To overcome this challenge, some countries choose to align their own monitoring and evaluation priorities with the reporting requirements attached to the external support. This, however, is problematic if the external reporting requirements do not meet domestic information needs. Alternatively, partner countries may choose to integrate some of the more general reporting requirements (e.g. on the allocation of resources and the implementation of planned activities) into national monitoring and evaluation systems already in place in many developing countries. This approach, however, does not address the issue of evaluation.

Challenges to the monitoring and evaluation of adaptation

Three methodological challenges affect the monitoring and evaluation of adaptation: i) measuring the attribution of adaptation interventions, ii) establishing baselines and setting targets in a relatively uncertain climate context, and iii) assessing long-term climate change adaptation (Dinshaw et al., 2014). While this is not a comprehensive list (see Bours, McGinn and Pringle, 2014), and none of the challenges are unique to climate change adaptation, their combined scope and scale are. Each challenge is briefly explored below.

Measuring attribution

The causal linkages between an intervention and change on the ground can be difficult to determine. This is particularly the case when countries take an integrated approach to adaptation. This means that adaptation considerations are integrated into all national planning and budgeting processes. As a result, adaptation is often a relatively small component of larger programmes, strategies and plans that may not explicitly target climate change but that nevertheless influence the country's climate resilience (e.g. disaster risk management and flood protection strategies). Such an integrated approach makes it difficult to determine the attribution[3] of specific initiatives to adaptation and to distinguish their impact from national development in general. However, if the strategic policy on adaptation is complemented by an action plan with clearly defined objectives, it may be possible to assess the attribution of these confined objectives in the short- and medium-term.

The underlying issue when measuring attribution is the lack of a "counterfactual" to assess what would have happened in the absence of a national approach to adaptation. Counterfactuals are usually established to facilitate the impact evaluation of an intervention by comparing the treatment group with a control group that closely resembles the treatment group but that did not benefit from the intervention (Gaarder and Annan, 2013). When focusing on a national approach to adaptation, however, it is not possible to distinguish between treatment and control groups.

To overcome the challenge of measuring attribution, the German monitoring and evaluation framework uses trend analysis to assess if climate impacts and vulnerabilities are changing over time (Schönthaler et al., 2010). Similarly, the proposed monitoring and evaluation framework for adaptation in Kenya's Climate Change Action Plan (KCCAP) examines climate vulnerability and institutional adaptive capacity (Republic of Kenya, 2012a). This approach examines the contribution of adaptation initiatives in keeping development on track (Brooks et al., 2011). In doing so, it uses bottom-up county-level indicators to assess the level of integration and capacity of climate risk management processes. At the same time, resilience outcomes and development performance are assessed using top-down national level indicators (see Chapter 4).

Establishing baselines and setting targets

It can be challenging to establish baselines for adaptation since national policies often do not include specific and measureable targets. Instead, developed countries commonly outline how the broader objective of reduced climate vulnerability and enhanced resilience may be achieved (Casado-Asensio and Steurer, 2013). Many developing countries seem to be taking a similar approach, outlining the overarching objectives (e.g. Kenya, Mozambique, Nepal) or sectoral action plans (e.g. the Philippines, see Chapter 5). Without clear and actionable targets, it can be difficult to track progress and to evaluate the attribution of a national approach to adaptation.

The challenge of setting targets is not unique to the context of adaptation. A review of monitoring and evaluation approaches of poverty reduction strategies found that "specifying clear targets, for which data are available, and identifying intermediate indicators remains particularly challenging" (IMF and World Bank, 2005). Further, the review suggests that many poverty reduction strategies would benefit from "a more explicit link between goals and targets and the policies needed to achieve them" (IMF and World Bank, 2005, 11). In an attempt to overcome this challenge, the Australian Government has proposed that the risks to essential services (e.g. energy and water supply) are clearly identified and that corresponding responsibilities are allocated to persons or organisations best placed to address the risks (Australian Government, 2013). This type of approach can provide a good basis for subsequently evaluating if the identified risks were the right ones and if they were adequately addressed by those in charge.

In the evaluation of development interventions, the baseline refers to the situation prior to an intervention (OECD 2002). In the context of climate change adaptation and mitigation, it has been argued that the use of a baseline as a comparator may be misleading since adaptation interventions will, by definition, take place in a changing environment with evolving climate-related hazards and risks (Brooks et al., 2011; Clapp and Prag, 2012). A more accurate assessment would, therefore, need to factor in these changing circumstances to establish a good understanding of what the situation would have been in the absence of a policy approach on adaptation (Brooks et al., 2011). For example, a simple before and after comparison may show that climate vulnerability has deteriorated, while a comparison to a counterfactual would reveal that the situation would have been even worse without the explicit and implicit adaptation initiatives in place. Interpretation of changes relative to the baseline should therefore account for climate change.

Addressing long time-horizons

The long time-horizons and the uncertain nature of climate change have implications for the monitoring and evaluation of adaptation interventions. Despite this, project and programme evaluations usually take place shortly after their completion. Depending on the nature of the initiative (e.g. drought or risk prevention), this may be years or decades before the impact of the intervention becomes apparent. Although the timing of national policy evaluations may be more flexible, political pressures would make it difficult to commit resources for adaptation evaluations on the basis that results will potentially only be known 20-30 years in the future. At the same time, the value of an evaluation and the lessons it generates may be lost if the evaluation is postponed too far into the future. One option to overcome this challenge is to focus assessments on the achievement of intermediate outcomes, through ongoing monitoring and real-time evaluation. This can help to ensure that learning continues, before the most severe climate effects manifest themselves.

In practice, the overwhelming response to this challenge has been for countries to identify a set of indicators that enables them to monitor changes in their adaptation priorities or objectives (GIZ, 2013). For example, Germany has developed an indicator set to monitor changes in the 15 action and cross-sectional fields[4] prioritised in the German Adaptation Strategy (Schönthaler et al., 2010; Schönthaler, Andrian-Werburg and Nickel, 2011). Similarly, the UK Climate Change Act 2008 requires that an independent assessment of progress made in implementing the National Adaptation Plan is presented to the UK Parliament two years after the publication of the Plan in 2013, and subsequently every two years (Great Britain, 2008).

Approaches to monitoring and evaluating adaptation

Many theoretical frameworks have in recent years been developed on how to monitor and evaluate adaptation (Ayers et al., 2012; Brooks et al., 2011; Frankel-Reed and Brooks 2008; GIZ, 2012; Pringle, 2011; PROVIA, 2013; Villanueva, 2011). The frameworks differ in their geographic focus, the intervention level, and the policy or programmatic orientation (Bours, McGinn and Pringle, 2013). To various degrees, the frameworks embody a theory of what successful adaptation entails and steps that can inform an assessment of whether agreed objectives have been achieved. Dedicated green and climate funds (e.g. the Adaptation Fund [2011], the Global Environmental Facility [2012], and the Climate Investment Funds [2012]) have also formulated monitoring and evaluation frameworks to assess the impact of their portfolio of activities. Similarly, the Green Climate Fund has developed an initial results management framework (2014). Although these fund level frameworks may not be directly applicable to national approaches to monitor and evaluate adaptation, they can inform partner countries' domestic frameworks. For example, in Mozambique, the results framework for the Pilot Program for Climate Resilience (PPCR), under the Climate Investment Funds, has been used as a basis for developing the national monitoring and evaluation framework for adaptation (IIED, 2013b).

Ford et al. (2013) have developed a typology of approaches to monitor and evaluate adaptation. The typology is global in scope, but can be tailored to the national level. It identifies two types of approach for monitoring adaptation: outcome based and systematic. Outcome-based evaluations examine the effectiveness of adaptation interventions in reducing the impacts from climate change. Systematic measures to monitor adaptation, on the other hand, rely on indicators or proxies to monitor and evaluate the status of adaptation over time. The typology (summarised in Table 1.1) includes four systematic approaches for monitoring adaptation (Ford et al., 2013):

- **Political readiness:** Examines countries' readiness to start adapting to climate change in terms of political leadership and the presence of key governance factors. These governance factors include institutional arrangements, stakeholder consultation, the availability of climate change information, the appropriate use of decision-making techniques, technology development and diffusion, and adaptation research.
- **Process-based approaches:** Assesses the processes through which adaptation initiatives are developed and implemented. This approach is mostly used for adaptation projects and programmes. When applied at the national level, it can entail the use of indicators to monitor policy development and implementation.
- **Policy and programme approaches:** Examines policy and programme approaches to characterise systematically the current state of adaptation at the national level. This can

Table 1.1. Typology of approaches for adaptation monitoring and evaluation

Category	Approach	Characteristics	Data sources	Strengths	Limitations
Outcome evaluation	**Outcome evaluation:** reduced negative climate change impacts	• Monitor climate-related losses, mortality, and morbidity, over time and in relation to adaptation • Examine impacts of climatic hazard events before and after adaptation	• Natural hazard loss database (e.g. emergency events database) • Public health data (e.g. mortality, morbidity, disease prevalence)	• Quantification of adaptation progress and effectiveness • Metrics can be monitored over time • Availability of standardised global datasets of hazard losses and mortality across regions • Legitimacy within policy evaluation community	• Applicable only where outcomes are directly observable • Difficulty of inferring causality between outcomes and adaptation • Potential for maladaptation not captured • Limited applicability to "soft" and mainstreamed adaptations • Long lead-time • Does not measure outcomes from adapting to wider (non-event-oriented) climate change
Systematic options for monitoring adaptation	**Adaptation readiness:** presence of key governance factors essential for effective and successful adaptation	• With regard to adaptation, evidence of: political leadership, institutional co-ordination, stakeholder involvement, availability of climate change information, appropriate use of decision-making techniques, consideration of barriers to adaptation, funding for adaptation, technology development and diffusion, and adaptation research	• Evidence of political leadership (e.g. attendance and speeches at climate change meetings, location of climate co-ordination unit within the government) • Amount of investment in adaptation research	• Not dependent on outcomes being visible • Captures readiness for future action and ability to effectively implement adaptation initiatives	• Need to validate how readiness translates to action • Limited availability of readiness metrics
	Process-based approaches: processes through which adaptation initiatives are developed and implemented in pursuance of a desired objective or outcomes	• Comparison of adaptation characteristics and steps of development to theoretically and empirically derived characteristics of adaptation success and best practice	• NAPA • Adaptation inventories	• Not dependent on outcomes being visible • Capture the key processes that are believed to underpin effective and successful adaptation	• Limited systematically collected data on process of adaptation development and implementation • Time intensive • Unproven link to adaptation success
	Analysing policies and programme approaches: monitoring and comparison of reported adaptation actions and their characteristics	• Analysis of characteristics of reported adaptation and comparison across regions, by vulnerability categories, over time, and with respect to adaptation objectives	• UNFCCC National Communications • NAPA • Adaptation inventories • National adaptation assessment	• Not dependent on outcomes being visible • Systematic and quantitative analysis of progress • Amenable for rapid assessment	• Success not directly measured • Results subject to reporting bias
	Examining measures of changing vulnerability: measurement of change in vulnerability in relation to adaptation	• Monitor aggregate vulnerability indices in relation to adaptation action • Focus on specific indicators which capture the generic determinants of vulnerability (e.g. access to education, poverty, health, and inequality) • Examine specific components of sensitivity and adaptive capacity to climate change impacts	• Climate Change Vulnerability Index • Environmental Sustainability Index • Global Climate Risk Index • Global Adaptation Index (GAIN)	• Not dependent on outcomes being visible • Readily available vulnerability indices globally • Amendable for rapid assessment	• Inability to capture determinants of vulnerability • Fundamental disagreement between indices on magnitude of vulnerability • Challenge of linking change in indices to adaptation

Source: Adapted from Ford, J.D. et al. (2013), "How to track adaptation to climate change: A typology of approaches for national-level application", *Ecology and Society*, 18(3). *http://dx.doi.org/10.5751/ES-05732-180340*.

include an initial stocktake of current actions with reference to the extent of adaptation taking place. Over time, the adaptation measures in place can be examined against stated objectives and identified adaptation needs.

- **Changing vulnerability:** Examines climate risk "hot spots" and predicts future vulnerabilities. This information can be used to inform adaptation planning. It can also provide a baseline against which adaptation can be monitored and evaluated. At the global level, vulnerability indices have been criticised for their inability to capture the dynamic process of climate vulnerability.

The objectives of the monitoring and evaluation framework will determine the most suitable approach. However, recognising that developing and implementing a national framework can be time-consuming and resource intensive, countries may choose to initially focus on aspects that can be monitored within existing limits of data availability and monitoring and evaluation capacity. Over time, the coverage and scope may gradually expand (GIZ, 2013). For monitoring and evaluation to contribute to learning, it is beneficial if they are based on demand for the information by those closely linked to policy-making processes. This includes annual budget negotiations and national planning processes.

A number of countries have, or are in the process of, developing domestic monitoring and evaluation frameworks. Table 1.2 provides an overview of some approaches being explored. The majority of the frameworks are still in the planning and development stage, with the exception of three countries where implementation has started (France, Norway, and the UK). Of these, Norway has emphasised that its approach is not a monitoring and evaluation framework in the traditional sense. Rather, Norway is using existing systems and initiatives to track adaptation and to continuously learn what approaches to adaptation are effective in reducing climate vulnerability and risk (GIZ, 2013). Some of the frameworks outlined in Table 1.2 specify desired outputs and outcomes (e.g. the Philippines and France). Others are closely aligned to, or informed by, major adaptation programmes (e.g. Nepal and Mozambique). A third group of countries have focussed their approach on monitoring changes in a number of priority areas (e.g. Germany and the UK).

Mapped against the typology proposed by Ford et al. (2013), the frameworks outlined in Table 1.2, generally fall into the second category of approaches that monitor adaptation. A few frameworks, nonetheless, do include an evaluative component (e.g. the Philippines, France, and the UK). For example, the objective of the Philippine's framework is to identify the approaches to adaptation that are most effective in bringing about the desired change and to understand how the change came about. To achieve this objective, the framework includes seven results chains reflecting the adaptation priorities identified in the National Climate Change Action Plan (NCCAP) 2011-2028. Each results chain outlines the ultimate, intermediate and immediate outcomes as well as activities, outputs and complementary indicators (see Chapter 5). The NCCAP specifies that although the plan includes long-term objectives, these are not fixed and can be adjusted if circumstances change (Philippines Climate Change Commission, 2011). To ensure that the plan remains relevant, it will be monitored on an annual basis and evaluated every three years. The annual monitoring will help prioritise adaptation needs and the allocation of budgets while the periodic evaluations will assess the efficiency, effectiveness and impact of the NCCAP (Philippines Climate Change Commission, 2011). These processes will inform government decision makers whether the approach is the right one, if circumstances are changing, and if adjustments in the plan or in the implementation mechanisms are needed.

Table 1.2. **Examples of national monitoring and evaluation frameworks for adaptation**

	Approach	Status
Australia	• Identifies risks to essential services (e.g. energy and water supply) and allocation of responsibilities to persons or organisations best placed to address the risks. • Indicators of adaptation drivers, activities and outcomes.	National Adaptation Assessment Framework under development, initial set of 12 indicators identified and currently subject of consultation. Under review.
Germany	• Climate change impacts and response indicators for 15 action and cross-sectional fields to monitor adaptation. • Periodic evaluation of the German Adaptation Strategy.	Indicator system under review. Reporting expected to start in 2015.
France	• Process indicators and some outcome indicators for 20 priority sectors.	Indicator system reflects the 230 measures identified in the French National Adaptation Plan 2011-2015. Operational.
Kenya	• Indicator-based system using outcome- and process-based indicators measured at national and county levels.	Monitoring, reporting and verification of actions under the Kenyan National Climate Change Action Plan, top-down and bottom-up indicators identified at the national and county level. System currently under review.
Morocco	• Using indicators to monitor changes in vulnerability, adaptation progress and their impacts. • Around 30 indicators in each of the two pilot regions.	Indicator system for the two regions integrated into the Regional Environmental Information System (SIRE). Under review.
Mozambique	• Monitor climate change impacts and inform national budget allocations/international climate finance.	Draft framework proposed, including a set of indicators. Under development. Full implementation expected by 2020.
Nepal	• Programme-level indicators (based on PPCR core indicators and indicators linked to NAPA priorities); matched by individual project-level indicators. • Qualitative documentation of lessons learned. • 149 sub-national "environmentally friendly" indicators for different sectors (including climate) and scales (household to district).	Indicator system piloted for eight climate change projects that form the core of Nepal's Climate Change Program. Under development.
Norway	• Process and impact monitoring using repeated surveys of exposure and adaptive capacity.	System focuses on learning by doing, structured around regular national vulnerability and adaptation assessments. Operational.
Philippines	• Indicators linked to results chains for seven strategic priority sectors. • Climate Change Vulnerability Indices for measuring, monitoring and evaluating local vulnerability and adaptation.	Preliminary set of mostly process indicators developed. Under review.
South Africa	• Established outcome-based system will be used to monitor climate change impacts at appropriate spatial density and frequency. • Report progress on the implementation of adaptation actions.	Preparatory phase. E.g. the monitoring and evaluation team is being assembled, South Africa's climate change actions are being mapped, the National Climate Change Response Database is being updated.
United Kingdom	• Mix of approaches: regular, detailed climate change vulnerability assessment; indicators to monitor changes in climate risks, uptake of adaptation actions and climate impacts; decision-making analysis to evaluate if degree of adaptation is sufficient to address current and future climate risks.	Regular, detailed adaptation assessments comprised of monitoring changes in climate risks using indicators, and evaluating preparedness for future climate change by analysing decision-making processes. Operational.

Source: Adapted from GIZ (2013), *Monitoring and Evaluating Adaptation at Aggregated Levels: A Comparative Analysis of Ten Systems*, Deutsche Gesellschaft für Internationale Zusammenarbeit, Eschborn.

Most national frameworks currently being developed include some elements of "systematic approaches for monitoring adaptation" (Table 1.1). In particular, climate change risk and vulnerability assessments are widely used to monitor if the identified risks change over time (e.g. Kenya, Morocco, Germany and the UK). Kenya's climate change action plan outlines a comprehensive list of potential and priority mitigation and adaptation needs (Republic of Kenya, 2012b). The complementary National Performance and Benefit Measurement Framework (NPBMF), referred to as the MRV+ system, tracks both mitigation and adaptation actions and the synergies between the two. Once the adaptation priorities for Kenya's NAP have been finalised, specific targets will be identified. These will inform the evaluative component of the MRV+ framework (Republic of Kenya, 2012a). In Morocco the monitoring and evaluation frameworks being established in two regions aim to assess changes in vulnerability in key sectors. These frameworks will also monitor the implementation of adaptation interventions with the goal of providing recommendations of possible adjustments when needed (GIZ, 2013).

Domestic circumstances, rather than theory, tend to determine the design of countries' monitoring and evaluation frameworks and their implementation. Further, the frameworks build to varying extents on monitoring and evaluation systems already in place. For example, the Kenyan MRV+ framework is aligned with the National Integrated Monitoring and Evaluation System that aims to improve management for development results (Republic of Kenya, 2012a). Similarly, in Nepal, all national projects and programmes are subject to standard progress reporting that informs the allocation of the national budget (IIED, 2013c). The nature of the monitoring and evaluation frameworks is also influenced by data availability. It is, therefore, common practice to do an initial survey to identify the information that is already collected on a regular basis (e.g. household surveys and standard financial reporting), information that will be collected in the future, and sources of information that could be adjusted to also capture relevant climate change information.

Notes

1. OECD DAC has since 1998 monitored development assistance targeting the objectives of the Rio Conventions through its Creditor Reporting System (CRS). The CRS differentiates between initiatives that target the Conventions as a "principal objective", a "significant objective", or not at all. Activities marked as having adaptation as a "principal" objective would not have been funded but for that objective; activities marked "significant" have other primary objectives but have been formulated or adjusted to help meet adaptation concerns (OECD, 2014).
2. Dedicated climate funds usually channel the money through multilateral development banks and/or development agencies. They will therefore in the reminder of this report fall under the broad category of "development co-operation providers".
3. Attribution is defined as here as "the ascription of a causal link between observed (or expected to be observed) changes and a specific intervention" (OECD, 2002).
4. The 13 action fields are: i) human health, ii) building sector, iii) water regime, water management, coastal and marine protection, iv) soil, v) biological diversity, vi) agriculture, vii) forestry and forest management, viii) fishery, ix) energy industry (conversion, transport and supply), x) financial services industry, xi) transport, transport infrastructure, xii) trade and industry, xii) tourism industry. The two cross-section fields are: xiv) spatial, regional and physical development planning, and xv) civil protection.

References

Adaptation Fund (2011), *Evaluation Framework*, Adaptation Fund Board, Bonn.

Australian Government (2013), *Climate Adaptation Outlook: A Proposed National Adaptation Assessment Framework*, Department of Industry, Innovation, Climate Change, Science, Research and Tertiary Education, Commonwealth of Australia, Canberra.

Ayers, J. et al. (2012), *CARE participatory monitoring, evaluation, reflection & learning (PMERL) for community-based adaptation (CBA)*, CARE International, Geneva.

Bours, D., C. McGinn and P. Pringle (2014), *Twelve reasons why climate change adaptation M&E is challenging*, SEA Change CoP, Phnom Penh and UKCIP, Oxford.

Bours, D., C. McGinn and P. Pringle (2013), *Monitoring and evaluation for climate change adaptation: A synthesis of tools, frameworks and approaches*, SEA Change CoP, Phnom Penh and UKCIP, Oxford.

Brooks, N. et al. (2011), "Tracking Adaptation and Measuring Development", *IIED Climate Change Working Paper*, No. 1, IIED, London.

Busan HLF-4 (Fourth High Level Forum on Aid Effectiveness) (2011), *Busan Partnership for Effective Development Co-operation*, HLF-4, Busan.

Casado-Asensio, J. and R. Steurer (2013), "Integrated strategies on sustainable development, climate change mitigation and adaptation in Western Europe: Communication rather than co-ordination", *Journal of Public Policy*, *http://dx.doi.org/10.1017/S0143814X13000287*.

CIF (Climate Investment Fund) (2012), *Revised PPCR Results Framework*, CIF, Washington, DC.

Clapp, C. and A. Prag (2012), "Projecting Emissions Baselines for National Climate Policy: Options for Guidance to Improve Transparency", *OECD/IEA Climate Change Expert Group Papers*, No. 2012/04, OECD Publishing, Paris, *http://dx.doi.org/10.1787/5k3tpsz58wvc-en*.

Dinshaw, A. et al. (2014), "Monitoring and Evaluation of Adaptation: Methodological Approaches", *OECD Environment Working Papers*, No. 74, OECD Publishing, Paris, *http://dx.doi.org/10.1787/5jxrclr0ntjd-en*.

Ford, J.D. et al. (2013), "How to track adaptation to climate change: A typology of approaches for national-level application", *Ecology and Society*, Vol. 18(3), *http://dx.doi.org/10.5751/ES-05732-180340*.

Frankel-Reed, J. and N. Brooks (2008), *Proposed Framework for Monitoring Adaptation to Climate Change*, Paper prepared for the GEF international workshop on evaluating climate change and development.

Gaarder, M. and J. Annan (2013), "Impact Evaluation of Conflict Prevention and Peacebuilding Interventions", *Policy Research Working Paper*, No. 6496, World Bank Independent Evaluation Group, Washington, DC.

GCF (Green Climate Fund) (2014), "Initial Results Management Framework of the Fund", available at: *http://gcfund.net/fileadmin/00_customer/documents/MOB201406-7th/GCF_B07_04_Initial_Results_Management_Framework_fin_20140509.pdf*.

GEF (Global Environment Facility) (2012), *Climate change adaptation – LDCF/SCCF adaptation monitoring and assessment tool*, GEF, Washington, DC.

GEF IEO (Independent Evaluation Office) (2013), "Adaptation to Climate Change", *Fifth Overall Performance Study (OPS5) of the GEF Technical Document*, No. 19, GEF IEO, Washington, DC.

GIZ (Deutsche Gesellschaft für Internationale Zusammenarbeit) (2013), *Monitoring and Evaluating Adaptation at Aggregated Levels: A Comparative Analysis of Ten Systems*, GIZ, Eschborn.

GIZ (2012), *Adaptation made to measure: A guidebook to the design and results-based monitoring of climate change adaptation projects*, GIZ, Eschborn.

Great Britain (2008), *Climate Change Act 2008*, The Stationary Office, London.

Handley, G. (2008), *Mutual Accountability at the Country Level: Mozambique Case Study*, ODI, London.

IEG (Independent Evaluation Group) (2013), *Adapting to Climate Change: Assessing World Bank Group Experience, Phase III of the World Bank Group and Climate Change*, IEG – World Bank, IFC, MIGA, Washington DC.

IIED (International Institute for Environment and Development) (2013a), *Tracking Adaptation and Measuring Development (TAMD) in Ghana, Kenya, Mozambique, Nepal and Pakistan: Meta-analysis findings from Appraisal and Design phase*, IIED, London.

IIED (2013b), *Tracking Adaptation and Measuring Development (TAMD) in Mozambique: Appraisal and Design Phase Report*, IIED, London.

IIED (2013c), *Tracking Adaptation and Measuring Development (TAMD) in Nepal: Appraisal and Design Phase Report*, IIED, London

IMF (International Monetary Fund) and World Bank (2005), *2005 Review of the Poverty Reduction Strategy Approach: Balancing Accountabilities and Scaling Up Results*, IMF and World Bank, Washington, DC.

IPCC (Intergovernmental Panel on Climate Change) (2012), *Managing the Risks of Extreme Events and Disasters to Advance Climate Change Adaptation: A Special Report of Working Groups I and II of the Intergovernmental Panel on Climate Change*, Field et al. (Eds.), Cambridge University Press, Cambridge and New York.

Kato, T. et al. (2014), "Scaling up and Replicating Effective Climate Finance Interventions", *Climate Change Expert Group*, No. 2014(1).

Mullan, M. et al. (2013), "National Adaptation Planning: Lessons from OECD Countries", *OECD Environment Working Papers*, No. 54, OECD Publishing, Paris, *http://dx.doi.org/10.1787/5k483jpfpsq1-en*.

OECD (Organisation for Economic Co-operation and Development) (2014), *OECD DAC Statistics: Aid to Climate Change Adaptation*, available at *www.oecd.org/dac/environment-development/Adaptation-related%20Aid%20Flyer%20-%20May%202014.pdf*.

OECD (2009), *Integrating Climate Change Adaptation into Development Co-operation: Policy Guidance*, OECD Publishing, Paris, *http://dx.doi.org/10.1787/9789264054950-en*.

OECD (2008), *Accra Agenda for Action*, OECD, Paris.

OECD (2005), *Paris Declaration on Aid Effectiveness*, OECD, Paris.

OECD (2002), *Glossary of Key Terms in Evaluation and Results Based Management*, OECD Publishing, Paris, *http://dx.doi.org/10.1787/9789264034921-en-fr*.

OECD (2001), *Evaluation Feedback for Effective Learning and Accountability*, OECD Publishing, Paris.

OECD (n.a.), *Mutual Accountability: Emerging Good Practice*, available at: *www.oecd.org/dac/effectiveness/49656340.pdf*.

OECD/UNDP (2014), *Making Development Co-operation More Effective: 2014 Progress Report*, OECD Publishing, Paris, *http://dx.doi.org/10.1787/9789264209305-en*.

Pringle, P. (2011), *AdaptME: Adaptation monitoring and evaluation*, UK Climate Impacts Programme, Oxford.

PROVIA (2013), *PROVIA Guidance on Assessing Vulnerability, Impacts and Adaptation to Climate Change*, Consultation document, UNEP, Nairobi.

Ranger, N., R. Muir-Wood and S. Priya (2009), *Assessing Extreme Climate Hazards and Options for Risk Mitigation and Adaptation in the Developing World*, World Development Report 2010 Background Note, World Bank, Washington, DC.

Republic of Kenya (2012a), *National Performance and Benefit Measurement Framework: Section B: Selecting and Monitoring Adaptation Indicators*, Ministry of Environment and Mineral Resources, Kenya.

Republic of Kenya (2012b), *National Performance and Benefit Measurement Framework: Section A: NPBMF and MRV+ System Design, Roadmap and Guidance*, Ministry of Environment and Mineral Resources, Kenya.

SADEV (Swedish Agency for Development Evaluation) (2012), *Mutual accountability in practice: The case of Mozambique*, SADEV, Karlstad.

Schönthaler, K., S. Andrian-Werburg and D. Nickel (2011), Entwicklung eines Indikatorensystems für die Deutsche Anpassungsstrategie an den Klimawandel (DAS), Dessau-Roßlau, UBA (updated in March 2014).

Schönthaler, K. et al. (2010), *Establishment of an Indicator Concept for the German Strategy on Adaptation to Climate Change*, available at: *www.uba.de/uba-info-medien-e/4031.html*.

UNDP (United Nations Development Programme) (2010), *Results-Based Management Handbook: Strengthening REM harmonization for improved development results*, UNDP, New York.

UNFCCC (United Nations Convention on Climate Change Conference) (2011), *Decisions adopted by the Conference of the Parties*, Report of the United Nations Framework Convention on Climate Change Conference of the Parties on its sixteenth session, held in Cancun 29 November-10 December 2010, available at: *http://unfccc.int/resource/docs/2010/cop16/eng/07a01.pdf*.

UNFCCC (n.a.), *National Adaptation Programmes of Action*, available at: *http://unfccc.int/adaptation/workstreams/national_adaptation_programmes_of_action/items/7567.php*.

Villanueva, P.S. (2011), "Learning to ADAPT: monitoring and evaluation approaches in climate change adaptation and disaster risk reduction – challenges, gaps and ways forward", *SCR Discussion Paper*, No. 9.

Wood, B. et al. (2011), *The Evaluation of the Paris Declaration, Final Report*, Danish Institute for International Studies, Copenhagen.

PART I

Chapter 2

Effective monitoring and evaluation of climate change adaptation

This chapter examines two important enabling factors for national monitoring and evaluation of adaptation: i) data availability and monitoring and evaluation capacity; and ii) good co-ordination between the providers and the users of climate information. It also explores how providers of development co-operation can support partner countries in putting in place these enabling factors.

Key messages

- A diverse set of environmental and socio-economic data that countries collect on a regular basis can inform the monitoring and evaluation of adaptation. Remaining data gaps can gradually be addressed by, for example, incorporating relevant adaptation questions into established data collection processes such as household surveys.
- Human and technical capacity are necessary for the monitoring and evaluation of adaptation. Capacity constraints can be difficult to overcome if financial and human resources are limited, or if monitoring and evaluation are not valued sources of information for national planning and budgeting processes. Changes in the incentive structure of public officials to use the findings from monitoring and evaluation can help overcome this challenge.
- Given the diverse set of data used to monitor and evaluate adaptation, a co-ordination mechanism can usefully link data producers and users. It is beneficial if such a co-ordination mechanism has the mandate and capacity to gather information across sectors and levels of decision-making (local, regional and national).
- Development co-operation providers can support the development of partner countries' own statistical systems by, to the extent possible, drawing on data collection mechanisms already in place for their own reporting requirements. When data gaps exist, development co-operation providers can support initiatives that will contribute to enhanced capacity of the partner country's statistical system rather than focus on the collection of data for discrete projects and programmes.

Data availability and human and technical capacity are important enabling factors for the monitoring and evaluation of adaptation. Most countries have in place a diverse set of monitoring and evaluation mechanisms within which a framework for adaptation can be situated. However, to ensure that the information generated from monitoring and evaluation contributes to adaptation planning and learning, a co-ordination mechanism can help to connect producers and users of the data. Further, a participatory process where stakeholders agree on the objectives of the monitoring and evaluation framework and contribute to shared procedures can create ownership and help ensure that the information generated is relevant for everyone involved. This, in turn, can increase the likelihood of the information subsequently being used in adaptation planning and budgeting processes. However, for this to happen, strong political leadership is important. Experience from monitoring and evaluation of poverty reduction strategies suggests that placing the institutional lead close to the centre of government will ensure greater authority of the monitoring and evaluation unit and create strong links to the policy planning and budgeting processes (Bedi et al., 2006).

Data availability and monitoring and evaluation capacity

Reliable time series of climate variables and other socio-economic indicators enable governments to detect, predict and respond to changes in climate risks and vulnerabilities and to monitor and evaluate the effectiveness of adaptation measures in place (WMO, 2007). Although an increasing number of developing countries are starting to report on climate variables in response to the need to monitor and evaluate their NAPA's and NAP's, the lack of good climate data continues to pose a challenge for many. While data availability is a particular challenge in countries that are in conflict or that have recently emerged from one, this challenge is not limited to fragile states. For example, when Brazil in 2010 audited the government's response to adaptation, auditors faced serious data constraints, in part due to limited access to meteorological data, data not being available in a digital format, and due to the absence of a centralised system to co-ordinate and store the data (see Box 3.7). Further, the information used to monitor and evaluate adaptation often comes from line ministries that in turn may rely on agencies and local governments to collect the information. Capacity and time constraints at the different levels can all affect the quality of the data.

The cost of collecting data, limited resources and the number of pressing development priorities, are some of the challenges countries face when trying to bring together a national database for adaptation. An approach sometimes used in the context of poverty reduction strategies is to perform a diagnosis of monitoring mechanisms already in place to get an overview of existing data availability. In the short-term, this can highlight what is possible to monitor and evaluate and how existing mechanisms can be rationalised to meet emerging needs. Examples include the termination of data collection activities no longer useful, the consolidation of activities carried out by more than one agency, or a reduction in the number of data platforms used (Bedi et al., 2006). In Niger, for example, such diagnosis found that there were 10 distinct databases and other government information systems in place. This resulted in the same data being collected by different agencies. At the same time, mixed data collection methodologies were used, preventing the harmonisation of different data sources (Bedi et al., 2006).

Given the relatively recent focus on monitoring and evaluation of adaptation, countries rely to a large extent on data collected for purposes other than adaptation. This includes social and economic data to for example monitor national development plans and other established indices such as the Human Development Index and the Millennium Development Goals. Common sources of data include household and living standard surveys, sectoral statistics, labour force reviews and so on. Experience to date, however, has shown that data collection processes often differ and there is a lack of alignment between global monitoring needs and national reporting capacities (Paris21, 2013).

When the data are not specifically tailored to the context of adaptation, it is beneficial if the capacity is in place to identify what information can be used and what new data needs to be generated to assess the country's climate vulnerability. The approach used by the UK for the 2012 national assessment of flood risk was to score datasets against a number of criteria to determine their statistical quality and relative strengths and weaknesses (see Box 2.1). However, such scorecard assessments can be difficult to do in practice if there is no central data repository or if there is limited co-ordination between data producers and users.

Over time, countries can gradually enhance their data availability by: i) collecting data using streamlined processes that assure a consistent quality and reporting format, ii) including sufficient detail in data collection efforts for adaptation to be characterised, and iii) by making the data collected available in a digital format (Ford et al., 2013). Since this may

Box 2.1. **Scoring of datasets for the UK's assessment of flood risk in 2012**

The criteria used to score datasets for the UK 2012 assessment on flood risk included:

- **Temporal coverage:** How many years of data are available?
- **Update frequency:** How often is the dataset updated?
- **Data measurement approach:** Has the data been measured through a monitor or a survey?
- **Availability:** Are the data publicly available or available for a fee?
- **Objectivity:** Is there a potential bias introduced as a result of the data collection procedure?
- **Statistical quality:** Do the statistical techniques employed conform to standard statistical procedures?
- **Relevance as an indicator:** How relevant is the dataset to a particular subject area?

The information for each criteria was recorded in a summary table. This enabled the discussion to move from having a list of ideal indicators to developing a set of indicators where data was available that met set standards.

Source: Harvey, A. et al. (2011), *Provision of research to identify indicators for the Adaptation Sub-Committee*, AEA, Edinburg.

not immediately be feasible in all developing countries, an alternative approach can be to include questions specific to climate change in established data collection processes, such as household surveys that in many cases are conducted every four to five years. Mozambique, for example, has included nine climate change questions in its household survey. The questions examine if households have suffered food, asset or income losses due to climate change, what their sources of information are on disaster and weather risks, approaches households have taken to minimise the impact from such shocks, and sources of support when they have suffered from climate change. Mozambique started data collection in 2014 and the initial results are expected to be available in December 2015 (INE, n.a.).

It may be necessary in some countries to identify new sources of data that can generate additional information needed to better understand the potential climate change risks. In such cases, existing mechanisms may be used to ensure that the data are grounded in national development objectives and contribute to the overall statistical plan, rather than respond to specific adaptation initiatives. Discrete project-level monitoring and evaluation can result in a concentration of domestic monitoring and evaluation capacity within non-state institutions (e.g. bilateral development agencies or other providers of support) that make up the majority of climate change actors in many developing countries (Bird, 2011; IIED, 2013a). Such project-based assessments can also undermine the sustainability of monitoring and evaluation mechanisms that rely on data with the right temporal and spatial scales.

To ensure the availability of data to monitor and evaluate adaptation, Nepal has established the Climate Change Knowledge Management Centre (NCCKMC). The objective of the centre is to generate, manage, exchange and disseminate relevant climate change information and capacity building services (IIED, 2013c). The NCCKMC was introduced in 2010, but is not yet operational. Similarly, Kenya has proposed a Climate Change Relevant Data Repository (CCRDR) to store and archive all data and information needed for the MRV+ framework (Republic of Kenya, 2012b). The repository will include: i) raw quality checked data; ii) processed data generated by technical analysis groups and other working groups; and iii) reports generated by the MRV+ system. It will benefit from data already collected by various ministries to monitor over 6 000 national indicators. Possible sources of information that the repository can draw upon are summarised in Table 2.1. The repository will complement other online systems already in place including the Electronics Projects Monitoring Systems (E-ProMIS) that monitors project implementation, and the Kenyan Environmental Information Network (KEIN). The operationalisation of such online systems, however, is resource intensive and often relies on data from district offices that may not have the capacity to use them. Experience to date has also been mixed. For example, while KEIN lacks funds for further development, it is estimated that out of 200 000 potential projects to be captured in the e-ProMIS system, only 1 500 have yet been added (IIED, 2013a).

Table 2.1. **Domestic sources of data to monitor and evaluate climate change adaptation in Kenya**

Data source	Relevant sector	Description of data
Kenya Meteorological Department (KMD)	All	KMD generates information from 36 synoptic stations, 3 upper air stations, over 3 000 volunteer rainfall stations, 4 marine tidal gauges, 24 automatic weather stations, 3 airport weather observing systems, 17 hydro-meteorological automatic weather stations, 4 lightning and thunderstorm detection systems and 3 satellite receiving stations.
KMD	Agriculture	KMD operates 14 agro-meteorological stations owned by KARI. In addition to climate data, the stations record data from the surrounding farms (e.g. crop variety, stage of development, damage by pests, disease and adverse weather, plant density and expected yield).
Kenya Agricultural Research Institute (KARI)	Agriculture Livestock	KARI collects data on e.g. food, horticultural and industrial crops, animal production, animal health, soil fertility, vegetation, agroforestry and irrigation. In the future KARI will also collect data on: climatic change risks, household vulnerability to climatic change in specific regions/production systems, performance of various crop varieties under different climatic conditions.
Department of Resource Surveys and Remote Sensing (DRSRS)	Forestry Wildlife Agriculture Livestock	DRSRS collects data on livestock/wildlife numbers and distribution; produces maps to monitor livestock/wildlife habitats, vegetation cover, forests, species composition, biofuel, biomass, crops, land degradation, and human settlements. It contributes to early warning systems for crop forecasting and to Land Information Management Systems from geospatial databases.
Water Resources Management Authority (WRMA)	Water	WRMA monitors flow volumes at 455 river gauging stations in the five major drainage basins. It has 17 hydro-meteorological automatic weather stations in the major water catchments for measuring surface discharge, used by the Kenya Energy Generating Company to monitor hydro-power generation under changing rainfall conditions.
Kenya Forest Service (KFS)	Forestry	KFS operates through 9 conservancies to provide national level statistics on forestry in general, forest cover and land use change, timber and fuelwood consumption patterns.
National Environment Management Authority (NEMA)	Water	The NEMA Geographic Information System laboratory focuses on water quality-monitoring, which can be used as an indicator of climate change.
Kenyan National Bureau of Statistics (KNBS)	All	KNBS holds socioeconomic data from the Population and Housing Census and associated surveys (e.g. the Welfare Monitoring Survey). These data cover e.g. gender, poverty, living conditions and occupation.
Ministry of State for Planning, National Development	All sub-sectors	The Medium Term Plan reports are rich in information that has relevance to all sub-sectors.
Monitoring and Evaluation Directorate (MED)	All	The annual Public Expenditure Reviews include process-based indicators that measure expenditure on adaptation and related activities. The reviews provide information on how public funds are being used and their impact.

Source: Adapted from Republic of Kenya (2012a), *National Performance and Benefit Measurement Framework: Section B: Selecting and Monitoring Adaptation Indicators*, Ministry of Environment and Mineral Resources, Kenya.

Approaches to data collection should be matched by capacities to use the data at the individual, organisational and system levels, as well as by demand for the data. At the individual level, there is a need for greater focus on acquiring the technical skills required to monitor indicators according to set standards. This also entails enhancing the capacity of stakeholders to interpret and use the data to inform national policy processes. Capacity building at the organisational and system levels is closely interlinked and refers to the presence of an institutional and legal infrastructure that supports the collection and reporting of data in a transparent manner (UNDP, 2009). Examples of capacity elements for each of the three levels are summarised in Table 2.2.

Table 2.2. **Examples of capacity building elements for monitoring and evaluation**

Level	Definition	Capacity elements
Individual personal level	The individual job performance and behaviours/actions of staff with monitoring and evaluation responsibilities	• Job requirements • Skill levels and needs • Performance reviews • Accountability and career progression • Access to information, training/re-training • Professional networking
Organisational level	The infrastructure and operations that need to be in place within each organisation to support the collection, verification and use of data for programme management and accountability	• Management process • Communication process • Human resource system and personnel structure • Financial resources • Information infrastructure • Organisational motivation
System level	The monitoring and evaluation functions across different organisations and how they interact, as well as the supportive policy and legal environment for monitoring and evaluation	• Policies, laws and regulatory actions that govern the collection and use of information • Resource generation and allocation for monitoring and evaluation • Systems for management and accountability • Resources, processes and activities across different organisations

Source: Adapted from USG (2007), *Building National HIV/AIDS Monitoring and Evaluation Capacity. A Practical Guide for Planning; Implementing; and Assessing Capacity Building of HIV/AIDS Monitoring and Evaluation Systems*, Office of the Global AIDS Coordinator, United States Government, Washington, DC.

In Ghana, the government assessed in 2012 the capacity of nine ministries, departments and agencies[1] to Manage for Development Results (MfDR). The objective of the assessment was to identify the strengths and weaknesses of the government's approach to MfDR. The assessment found that there was a significant lack of capacity to monitor and evaluate public policies in all sectors. Further, the majority of sectors lacked the capacity to analyse statistical data and to use monitoring and evaluation findings to inform decision-making processes (Government of Ghana, 2012). To ensure evidence-based decision-making processes, the assessment concluded that relevant government ministries, departments and agencies would have to enhance their capacity to MfDR.

Such capacity constraints, however, can be difficult to overcome since resource scarcity often means that staff time is not dedicated to monitoring and evaluation. An additional challenge many developing countries face is the high turn-over of staff. In Nepal, a review found that officials usually do not receive formal training when taking up their roles as adaptation monitoring and evaluation officers, and often transfer to positions considered more prestigious as soon as they have acquired the necessary skills (IIED, 2013c). Furthermore, government officials account in some cases for a relatively small proportion of climate change actors. In Kenya, for example, it is estimated that non-state institutions account for over 70% of climate change actors (IIED, 2013a). Unless monitoring and evaluation become valued sources of information, such barriers are likely to persist. An approach governments may consider is changing the incentive structure of public officials to use the findings from monitoring and evaluation in their planning and budgeting processes and in their accountability structures (Mackay, 2007).

The potential role of development co-operation providers

The data constraints countries face in the context of adaptation are similar to the constraints they face when monitoring and evaluating other development priorities. Lessons learned from development practice can, therefore, inform development support targeted at enhancing data availability for adaptation. For example, experience has shown that efforts to enhance data availability are more likely to succeed and be sustained if they fit within the broader national strategy for the country's statistical system. This refers to the entire network of providers of data and other information. Further, to avoid that domestic resources get skewed towards the collection of data for externally financed programmes, providers of development co-operation may consider more flexible mechanisms for supporting statistical institutes in partner countries (Bedi et al., 2006). Lessons from poverty reduction strategies are summarised in Box 2.2.

Box 2.2. **Challenges to monitoring and evaluating poverty reduction strategies**

Since the early 1990s, development co-operation providers have supported a number of developing countries in enhancing the capacity of their national statistical systems in monitoring and evaluating national poverty reduction strategies. Despite this support, a number of challenges remain. Many of these are relevant to the monitoring and evaluation of adaptation:

- A number of countries have developed statistical master plans and established inter-institutional committees responsible for linking national statistics institutes with data users. In many countries, the statistical master plans predate the monitoring and evaluation systems introduced for poverty reduction strategies, and have not subsequently been revised. This has resulted in overlapping co-ordination structures and redundant committees.
- National statistics institutes tend to prioritise large surveys and other statistical operations for which financial support from development agencies is available. For example, only one-fifth of support from development co-operation providers to Malawi's statistical system went to regular statistical activities; the remaining four-fifths went to irregular project and programme activities.
- Large surveys are used to monitor poverty reduction strategies. Although national statistics institutes often offer training on how to use of the data, agencies often prefer to use their own data, although it may not be of comparable quality.
- National statistical systems refer to both central statistics agencies and other producers of data. However, in many countries there is a disconnect between central agencies and the wider system, resulting in data gaps and redundancies.

Source: Bedi, T. et al. (2006), *Beyond the Numbers: Understanding the Institutions for Monitoring Poverty Reduction Strategies*, World Bank, Washington, DC.

Providers of development co-operation create demand for data through the results frameworks that partner countries have to report on, but they can also support the production of that data (e.g. the collection of meteorological data). Further, development support can assist partner countries in analysing existing data collection and sharing mechanisms, assessing the availability of relevant data, and in identifying further information needs. Once national statistics strategies have been developed and data collection systems established, development co-operation providers can play an important role in ensuring their sustainability (Paris21, 2007). This may entail greater support to partner countries for administrative functions rather than for the implementation of particular activities or surveys (Bedi et al., 2006).

Providers of development co-operation will also play an important role in enhancing the capacity of partner countries to use relevant data to better understand the links between climate change risks, vulnerabilities and resilience, and to use that information to inform domestic planning and budgeting processes. Support for capacity building initiatives can help to ensure that data users are in a good position to infer policy implications based on documented risks and impacts. At the individual level, capacity building support can promote learning, critical thinking, team building and action planning. At the organisational and system levels, it can contribute to an environment that is open to self-reflection and learning. To ensure sustainability, such capacity building initiatives benefit from a good understanding of the local context and partner countries' own priorities. This does not entail a simple transfer of skills but rather sustained support over a period of time.

Co-ordination between providers and users of climate information

The tools or sources of information presented in Chapter 3 cover a number of activities undertaken by different government and non-government agencies and institutions. While monitoring may be an integral component of the design and implementation of an adaptation policy, it does not capture progress made in implementing other initiatives that contribute to the country's climate resilience. Similarly, evaluation mechanisms assessing the efficiency and cost-effectiveness of adaptation policies and other relevant initiatives may be centrally managed and external to the daily management of implementing the policy. If no mechanism is in place to ensure that findings from the two processes are readily available to each other, good co-ordination between producers and users of the various sources of information is useful.

A co-ordination unit can be situated within the institutional mechanisms already in place for the adaptation planning process or within the body responsible for the monitoring and evaluation of a country's development priorities. Alternatively, an independent body can be established to co-ordinate the monitoring and evaluation process. Using existing institutional mechanisms can be effective in reducing the risk of duplicating efforts, while the creation of an independent body can signal the importance attributed to the monitoring and evaluation of adaptation and ensure a degree of independence. It is important that the system chosen has the mandate and the capacity to gather information across sectors and adaptation priority areas, as well as across different levels of decision-making (local, regional and national). This will ensure that it is in a position to assess progress made on adaptation and to identify remaining gaps and challenges (GIZ, 2013).

Depending on the nature of the system, the role of the co-ordination unit can be to collect as well as to analyse relevant data. This approach is used by the UK Adaptation Sub-Committee (ASC). The ASC works with sectoral experts to determine what aspects of adaptation to focus on within pre-defined thematic areas. Drawing on different sources of data that are publicly available or available free of charge, the ASC assesses the UK's preparedness to face identified climatic risks (ASC, 2012). Alternatively, the co-ordination unit can task the organisations owning the data to report on a pre-defined set of indicators. This is the approach used by Germany (Schönthaler et al., 2010). A third approach proposed by Kenya is to create a central online data repository where all relevant data and information is stored. This process will be supported by the Data Supply and Reporting Obligation Agreements to ensure that all relevant stakeholders report their data (Republic of Kenya, 2012a).

To facilitate good monitoring and evaluation of adaptation, it is beneficial to engage the co-ordination unit in the development and implementation of the monitoring and evaluation plan from the outset. This ensures that all stakeholders are clear on their respective roles and responsibilities and that there is a good understanding of the various sources of information. In Mozambique, for example, the national approach to adaptation is situated within the National Development Strategy to ensure a climate resilient future. As part of the Development Strategy, each sector is responsible for identifying individual indicators and targets on adaptation (IIED, 2013b). In order for a co-ordinating unit to be in a position to draw on this information it would ideally be connected to the national planning process from the outset and collaborate closely with key stakeholders.

Further, good co-ordination is important given the time it takes to implement a monitoring and evaluation system. To illustrate, Mozambique initiated work on its monitoring and evaluation framework in 2012 when the National Strategy for Climate Change was introduced. The framework will be developed on an incremental basis, building on lessons learned. The initial report will be submitted to the Council of Ministers in 2015, but the system will not be fully operational until 2020 (IIED, 2013b). Similarly, the statutory duty of the UK ASC was identified in the UK Climate Change Act 2008, but the first report to the parliament on the UK National Adaptation Programme is scheduled for 2015 (ASC, 2011).

The potential role of development co-operation providers

A co-ordinated monitoring and evaluation process ought to be domestically owned and led. This ensures that the results are useful for domestic policy-makers, while also making it more likely that the process will be sustained over time. Development co-operation providers, however, can play an important role in facilitating the co-ordination process through financial and technical support. In countries without an existing domestic monitoring and evaluation framework for adaptation, providers of development co-operation can help to identify what systems are already in place that a framework for adaptation can build upon. Development co-operation providers can also facilitate knowledge sharing among developing countries and invest in identifying promising practices that can potentially be scaled up elsewhere.

Development co-operation providers can also support partner countries by, to the extent possible, aligning their own monitoring and evaluation efforts with domestic systems. Further, they can ensure that the results from their monitoring and evaluation efforts are available to the national co-ordination unit. This however, may be difficult in practice. Experience with the Performance Assessment Framework in Mozambique (discussed in Box 1.3) demonstrates that despite a formal agreement by the government and the supporting development agencies to use the assessment framework jointly agreed upon, the co-ordinated approach has been challenged by the prevalence of stand-alone projects and programmes. With the arrival of climate change-related funding, different actors have also tried to position themselves as being best placed to access the additional climate funds (IIED, 2013b). It may however, also be linked to the fact that large adaptation initiatives often come with their own reporting frameworks.

Note

1. Ministry of Education, Ministry of Finance and Economic Planning, Ghana Statistical Service, Ministry of Health, Ministry of Local Government and Rural Development, Ministry of Roads and Highways, Ministry of Women and Children's Affairs, Ministry of Food and Agriculture, and the National Development Planning Commission.

References

ASC (UK Adaptation Sub-Committee) (2012), *Climate change – is the UK preparing for flooding and water scarcity?*, ASC, Committee on Climate Change, London.

ASC (2011), *Adapting to climate change in the UK: Measuring progress*, ASC, Committee on Climate Change, London.

Bedi, T. et al. (2006), *Beyond the Numbers: Understanding the Institutions for Monitoring Poverty Reduction Strategies*, World Bank, Washington, DC.

Bird, N. (2011), *The Future for Climate Finance in Nepal*, Overseas Development Institute, London.

Ford, J.D. et al. (2013), "How to track adaptation to climate change: A typology of approaches for national-level application", *Ecology and Society*, Volume 18(3), *http://dx.doi.org/10.5751/ES-05732-180340*.

GIZ (Deutsche Gesellschaft für Internationale Zusammenarbeit) (2013), *Monitoring and Evaluating Adaptation at Aggregated Levels: A Comparative Analysis of Ten Systems*, GIZ, Eschborn.

Government of Ghana (2012), *Capacity Assessment for Effective Delivery of Development Results in Ghana*, Government of Ghana, Accra.

Harvey, A. et al. (2011), *Provision of research to identify indicators for the Adaptation Sub-Committee*, AEA, Edinburg.

IIED (International Institute for Environment and Development) (2013a), *Tracking Adaptation and Measuring Development (TAMD) in Kenya: Appraisal and Design Phase Report*, IIED, London.

IIED (2013b), *Tracking Adaptation and Measuring Development (TAMD) in Mozambique: Appraisal and Design Phase Report*, IIED, London.

IIED (2013c), *Tracking Adaptation and Measuring Development (TAMD) in Nepal: Appraisal and Design Phase Report*, IIED, London.

INE (Instituto Nacional de Estatística de Moçambique) (n.a), Climate change questions added to the Mozambique Household Survey, Personal communication with the Instituto Nacional de Estatística de Moçambique, Maputo.

Mackay, K. (2007), *How to Build M&E Systems to Support Better Government*, World Bank Independent Evaluation Group, Washington, DC.

Paris21 (2013), "Engineering a Development Data Revolution: 2013 UN General Assembly Side Event", available at: *www.paris21.org/node/1593*.

Republic of Kenya (2012a), *National Performance and Benefit Measurement Framework: Section B: Selecting and Monitoring Adaptation Indicators*, Ministry of Environment and Mineral Resources, Kenya.

Republic of Kenya (2012b), *National Performance and Benefit Measurement Framework: Section A: NPBMF and MRV+ System Design, Roadmap and Guidance*, Ministry of Environment and Mineral Resources, Kenya.

UNDP (United Nations Development Programme) (2009), *Handbook on Planning, Monitoring and Evaluation for Development Results*, UNDP, New York.

USG (United States Government) (2007), *Building National HIV/AIDS Monitoring and Evaluation Capacity. A Practical Guide for Planning; Implementing; and Assessing Capacity Building of HIV/AIDS Monitoring and Evaluation Systems*, Office of the Global AIDS Coordinator, USG, Washington, DC.

WMO (World Meteorological Organisation) (2007), *WMO's Role in Global Climate Change Issues with a focus on Development and Science-based Decision Making: Position Paper*, WMO, Geneva.

National Climate Change Adaptation
Emerging Practices in Monitoring and Evaluation

PART I

Chapter 3

National tools for monitoring and evaluation of climate change adaptation

This chapter identifies four tools or sources of information that countries may consider when monitoring and evaluating adaptation: i) climate change risk and vulnerability assessments, ii) indicators to monitor prioritised adaptation needs; iii) lessons learned from adaptation initiatives, and iv) national audits and climate expenditure reviews. For each tool, the potential role of development co-operation providers in supporting partner countries is discussed.

Key messages

- The broad nature of adaptation demands a portfolio of monitoring and evaluation tools that when combined provide an overview of the larger resilience picture. The composition of the tools used will be most effective if they reflect domestic circumstances and capacities.
- Climate change risk and vulnerability assessments can provide a baseline of domestic vulnerabilities to climate change against which progress on adaptation can be reviewed. If repeated, such assessments can also demonstrate how risks and vulnerabilities are changing over time.
- Indicators facilitate an assessment of progress made in addressing adaptation priorities. On their own, however, indicators cannot explain how the change came about. Reporting on, and using indicators, is resource intensive. They must therefore be carefully defined, and when possible, draw on existing data sources.
- Project and programme evaluations can help to identify what approaches to adaptation are effective in achieving agreed adaptation objectives and to understand what some of their enabling factors for success may be.
- National audits and climate expenditure reviews examine if resources allocated for adaptation are appropriately targeted and allocated cost-effectively. This information may be particularly useful when resources are specifically earmarked for adaptation.
- Development co-operations can provide technical support to partner countries implementing monitoring and evaluation tools. To ensure a sustainable approach that contributes to domestic systems already in place, co-ordination and commitment to support partner countries beyond the initial implementation phase is ideal. Development co-operation providers can also play an important role in facilitating peer learning and the exchange of lessons learned.

This chapter identifies four tools or sources of information that can provide a basis for efforts to develop a national framework for adaptation. These tools build on existing approaches currently being tested in different country contexts. All four tools may not be relevant for every country context and their applicability may also change over time. It is therefore important to build in the flexibility to respond to changing adaptation needs and to ensure that the national approach to adaptation reflects the state of climate science and builds on lessons learned. For each tool, the potential role of development co-operation providers in supporting partner countries is discussed.

Climate change risk and vulnerability assessments

Climate change risk and vulnerability assessments can help to identify priority adaptation needs; when repeated they can illustrate how these priorities are changing over time. The role of risk and vulnerability assessments in adaptation planning is emphasised in the UNFCCC technical guidelines for the preparations of NAPAs, NAPs and National

Communications. All three processes encourage countries to assess what the adverse impacts from climate change may be, where risks are projected to increase, what the priority adaptation needs are, and how these can be addressed taking into account the projected magnitude, probability, and urgency of the risk (LEG, 2012; LEG, 2002; UNFCCC, 2000). When national climate change risk and vulnerability assessments are not available, relevant information can be derived from sub-national or programme-level assessments. Techniques are available that enable the users of the information to normalise the scales so that separate risk assessments can be aggregated at the national level (GIZ, 2014a).

Climate change risk and vulnerability assessments are first and foremost a tool used to identify key vulnerabilities. This information is often used to guide the allocation of resources to priority adaptation needs. Once the climate risks and their likelihood have been established (e.g. on a scale ranging from "almost certain" to "rare"), a priority rating can be assigned based on the projected consequences of the risk (e.g. ranging from "catastrophic" to "insignificant") (Australian Government, 2006). Alternatively, risk assessments can focus on understanding the risks, followed by assessments on how to target priority risks and manage residual risks (OECD, 2013). While some events can happen on a recurring basis (e.g. structural damages or agricultural losses), others are likely to happen only once (e.g. the loss of endangered species or the relocation of vulnerable populations).

Table 3.1. **Prioritising climate change risks**

Likelihood	Consequences				
	Insignificant	Minor	Moderate	Major	Catastrophic
Almost certain	*Medium*	*Medium*	*High*	*Extreme*	*Extreme*
Likely	*Low*	*Medium*	*High*	*High*	*Extreme*
Possible	*Low*	*Medium*	*Medium*	*High*	*High*
Unlikely	*Low*	*Low*	*Medium*	*Medium*	*Medium*
Rare	*Low*	*Low*	*Low*	*Low*	*Medium*

Source: Australian Government (2006), *Climate Change Impacts and Risk Management: A Guide for Business and Government*, Australian Greenhouse Office, Department of the Environment and Heritage, Canberra.

Stakeholder participation, reflecting the breadth of adaptation policy-making, is important to ensure that climate change risk and vulnerability assessments produce policy relevant information. This contributes to an increased awareness of the risks but also a sense of ownership of the process and the adaptation options that get proposed as a result of it (GIZ, 2014a; Cardona et al., 2012). Table 3.2 outlines a few examples of information that may be considered when conducting a risk and vulnerability assessment.

Climate change risk and vulnerability assessments can also be a useful tool to monitor how adaptation priorities are changing over time. They provide a basis against which subsequent changes in the country's adaptation priorities can be assessed. When the assessments are repeated on a regular basis, this provides periodic "snapshots" of the adaptation priorities and the emerging priority risks and vulnerabilities. To understand the underlying drivers of the changing priorities, the risk and vulnerability assessments need to be matched by additional context analyses (GIZ, 2014a). Climate change risk and vulnerability assessments can themselves also be evaluated to examine their success at identifying the relevant adaptation priorities. Further, they can contribute to the evaluation of the effectiveness and relevance of the policy approach on adaptation.

Table 3.2. **Examples of data and information that can be used in risk and vulnerability assessments**

Vulnerability aspect	Examples of relevant information
Hazard: Potentially damaging climate influence that may adversely affect a valued attribute of a system at the national and local level	• Quantitative models that project precipitation and temperature changes at different scales • Quantitative models that examine the consequences of temperature and precipitation changes (e.g. drought, flood, sea level rise, changes in pest and disease outbreaks) • Qualitative information (e.g. expert judgment and stakeholder consultations), that can enhance or validate information about local-level climate hazards
Exposure: The presence of people and assets in areas that could be adversely affected by climate hazards	• Hazard maps depicting the location and distribution of people, infrastructure and ecosystems in areas that are or may be affected by hazards
Sensitivity: The degree to which people and assets are affected, positively or negatively, by climate variability or change	• Database of previous impacts of hazards – e.g. crop loss, economic loss, human and animal deaths • Models to estimate the impact of past or future climate hazards on e.g. crops, livestock and ecosystems • Maps depicting the location and distribution of fragile or poor quality housing, land, infrastructure, as well as degraded ecosystem and marginal populations • Local observations, experiences with climate hazards
Adaptive Capacity: The general ability of institutions, systems, and individuals to adjust to potential damage, to take advantage of opportunities, or to cope with the consequences	• Development data and indices (e.g. population, inequality, debt, economic productivity, trade flows, education levels, foreign direct investment, disease patterns) • Ecosystem goods and services • Census data, household surveys • Institutional capacity assessments • Local coping and adaptation strategies

Source: Adapted from GIZ (2013c), *Comparative analysis of climate change vulnerability assessments: Lessons from Tunisia and Indonesia*, Gesellschaft für Internationale Zusammenarbeit, Eschborn.

The UK's Climate Change Risk Assessment (CCRA) provides an example of how such assessments can contribute to monitoring and evaluation of adaptation. The first CCRA was produced in 2012, in preparation for the publication of the National Adaptation Programme in 2013. The CCRA provides a baseline of projected climate risks in the absence of current or planned action. It will be repeated every five years as mandated by the Climate Change Act 2008. The five-yearly cycle is intended to ensure that the most pressing climate risks are continuously assessed (Defra, 2012). Over time, this could serve as a form of monitoring where changes in the prioritised risks or the magnitude of existing risks can provide an overview of UK's vulnerability to climate risks and how it changes over time (see the methodology summarised in Box 3.1).

Kenya also undertook a risk and vulnerability assessment during the preparatory stage of the KCCAP. This assessment helped the government identify, prioritise and rank the most important climate risks. For each of the risks, the potential climate impacts were examined within the context of Kenya's development needs. This assessment was based on a literature review, stakeholder consultations, technology needs assessments and a review of relevant national planning documents (Republic of Kenya, 2012). Where possible, the climate risks were also assessed in terms of their expected economic costs. Despite the wealth of information gathered in the risk assessment, it is not referred to in Kenya's proposed MRV+ framework.

Although the important role of climate change risk and vulnerability assessments for adaptation planning is well understood, examples to date have been variable in terms of their breadth of coverage. Two independent reviews of international and European climate change planning[1] found that the majority of governments examined had yet to undertake comprehensive assessments. Assessments that had been conducted tended to have a sectoral focus or to be based on a climate scenario of a 2°C temperature increase, rather

Box 3.1. **The risk assessment methodology used for the UK's Climate Change Risk Assessment**

The assessment methodology used to identify and prioritise climate risks and opportunities in the UK consisted of five elements:

- **Risk screening:** Literature review and consultation in 11 research areas. This resulted in a list of more than 700 potential climate change risks;
- **Risk selection:** A scoring exercise that considered the magnitude and likelihood of risks and the perceived urgency of adaptation action. This reduced the list to 100 risks categorised into five themes: i) agriculture and forestry, ii) business, iii) health and wellbeing, iv) buildings and infrastructure, and v) natural environment;
- **Assessment of vulnerability:** Further research on other non-climate factors that influence future risks (e.g. social vulnerability of society or institutional capacity to respond to future climate change risks);
- **Evaluation of current risks:** An evaluation of current risks drawing on best available information from government departments and the regulated industries;
- **Assessment of future risks:** An examination of the sensitivity of identified risks to climate variables, considering the effect of future climate change and variability on the current population. Within the context of projected population changes, total climate risks for future time periods were categorised as "high", "medium" and "low".

The five elements are, in theory, transferable to different country contexts, but the availability of data and domestic capacity may be a barrier in some countries. The potential role of development co-operation providers in supporting partner countries in overcoming these barriers was discussed in Chapter 2.

Source: Defra (2012), *The UK Climate Change Risk Assessment 2012: Evidence Report*, Defra, London.

than a range of possible temperature increases (INTOSAI, 2010a; EUROSAI, 2012). A possible explanation is that climate change risk and vulnerability assessments are time and resource intensive. However, it is possible that this trend will change in the future as many developing countries, in the context of their NAPAs, NAPs and National Communications have received support to undertake risk and vulnerability assessments.

The potential role of development co-operation providers

Development co-operation providers currently play an important role in supporting partner countries in establishing the data collection mechanisms needed to conduct climate change risk and vulnerability assessments. Such support can entail a detailed mapping of data that are already available, other information that can easily be collected, and data that are not yet available but will be needed in the future. In building on these efforts, development co-operation providers can extend their support to the production of partner countries' risk and vulnerability assessments conducted for their NAPAs, NAPs and related national planning processes. Providers of development co-operation can also draw on partner countries' climate assessments to inform their own strategies and co-operation programmes in ways that effectively target priority areas for adaptation.

To ensure that risk and vulnerability assessment tools meet the needs of the intended users, Hammil and Tanner (2011) have put forward a number of recommendations that development agencies may wish to consider as they plan their support:

- Support capacity building initiatives, including training on how to conduct a climate change risk and vulnerability assessment.
- Strengthen the links between groups that generate climate information and those that use the information for policy planning processes to ensure a better match between the availability and need of climate data.
- Improve guidance and support that will enable national authorities to use the data for adaptation planning through the development of common guidance and enhanced stakeholder engagement.
- Contribute to a harmonised approach to risk and vulnerability assessment that can be tailored to specific contexts and needs based on a common terminology.
- Work closely with partner countries to ensure ownership and the use of risk and vulnerability assessments in national adaptation planning processes, and development approaches in general.

Indicators for monitoring progress against adaptation priorities

National adaptation planning processes often set out the strategic direction on adaptation without specifying outcomes and targets (GIZ, 2013a; Casado-Asensio and Steurer, 2013). In this case, the monitoring and evaluation framework may need to elaborate indicators reflecting those priorities. These can then provide a useful tool for tracking progress in adapting to climate change and informing subsequent policy planning and budgeting processes. A perennial issue with indicators is that they may be skewed towards issues that can easily be measured and where data are available, rather than issues of particular interest. At the national level, there is an underlying tension to be managed between three objectives: ensuring sufficient stability of the indicator set to allow comparisons over time, retaining a manageable number of indicators, and having the flexibility to respond to changing priorities.

The collection and use of national indicators is resource intensive. It is, therefore, important that the indicators facilitate comparison across geographic scales and over time. Further, when defining an indicator set for adaptation, broad stakeholder consultation can help target the information generated towards prioritised adaptation needs and ensure that it addresses existing information gaps. Stakeholder consultation starts with the national (and in some cases local) authorities responsible for adaptation and climate change. Sectoral experts knowledgeable about the projected climate change risks and vulnerabilities can play an important role in identifying suitable indicators. Further, collectors and holders of national data understand what aspects of adaptation can be measured, what historical data are available, and what information is likely to be collected in the future. The contribution of data collectors to such consultations, however, will be contingent on them having a basic understanding of adaptation.

This consultative approach to indicator development and the alignment of indicators to the national adaptation priorities is reflected in emerging practice. For example, Kenya's proposed MRV+ framework will, to the extent possible, draw on indicators already being monitored for the country's Vision 2030 strategy and for related national, sub-national and sectoral plans and strategies (Republic of Kenya, 2012). To identify a suitable indicator set to monitor progress, the government applied a methodology developed by the International Institute on Environment and Development (IIED) called Tracking Adaptation and Measuring Development (TAMD) (summarised in Box 3.2). Through a consultative

Box 3.2. Tracking Adaptation and Measuring Development

The International Institute for Environment and Development (IIED), in partnership with Garama 3C Ltd. and Adaptify, and supported by the UK Department for International Development has developed a framework for monitoring and evaluating adaptation called Tracking Adaptation and Measuring Development (TAMD). The framework proposes a two-track process: i) an "upstream" assessment of the level of integration and capacity of climate risk management processes, and ii) a "downstream" assessment of resilience outcomes and development performance in the context of climate change. The objective of the TAMD framework is to assess the effectiveness of adaptation efforts in keeping development on track whether that be through a national system, a project, or understanding the contribution of a set of interventions to building resilience.

The framework and the selection of indicators is tailored to the purpose of the evaluation and the specific hazard and development context. The indicators can be grouped into three categories:

- Indicators to assess the extent and quality of climate risk management.
- Development and resilience indicators that measure whether development is on track.
- Contextual indicators on climate hazards and the external environment.

IIED is working with several national and sub-national governments to develop bespoke evaluation frameworks at different scales including in Kenya, Nepal, Pakistan, Mozambique, Ethiopia and Cambodia. In Kenya for example, TAMD is being used to assess and strengthen the performance of county level Climate Adaptation Funds.

Source: IIED (2013), *Tracking Adaptation and Measuring Development (TAMD) in Ghana, Kenya, Mozambique, Nepal and Pakistan: Meta-analysis findings from Appraisal and Design phase*, International Institute for Environment and Development, London.

process, 20 national and county-level indicators have been identified, all of which link to existing measures regularly assessed at the national or county level (see the full list of indicators in Chapter 4). These indicators are supported by a larger set of process-based output indicators.

In the Philippines, the proposed output and outcome indicators are aligned with the seven strategic priorities outlined in the NCCAP[2] (see Chapter 5). The indicators build to the extent possible, on data already collected for national, local and sectoral development strategies and plans. However, the proposed framework also considers what additional indicators may be needed to fulfil the monitoring and evaluation requirement of the NCCAP (GIZ, 2013a). Climate Change Vulnerability Indices will be developed to complement the indicators already identified. These indices will provide a set of standard indicators consistent with the seven thematic areas and will be applicable to all climate change initiatives at the national and sub-national level. The objective of these indices is to streamline climate change initiatives and the complementary collection of information to better facilitate comparative assessments and the exchange of lessons learned (GIZ, 2013a).

Similarly, the UK monitoring and evaluation framework aims to measure the preparedness of the society and the economy to the projected impacts of climate change. The decision to focus on climate preparedness, rather than on climate impacts and responses, is based on the need to acquire long time series to effectively measure impacts. The 2012 assessment examining the risk of flooding was, therefore, based on a set of indicators that prioritised a relatively small set of the impacts identified in an initial system mapping. This illustrates the point that it is neither possible nor desirable to measure every possible impact (the indicators used for the 2012 and 2013 assessments are summarised in Chapter 6). In order to identify the most significant impacts and risks, an evidence-based assessment was undertaken. This assessment was based on three criteria that were matched by a review of the availability, relevance and quality of existing datasets suitable for the proposed indicators (Harvey et al., 2011):

- The *significance* of the impact to the UK society, environment and economy, focusing on the current situation.
- The *sensitivity* of the impact to climate.
- The expected *future changes* in impact anticipated under climate projections.

The UK approach of focusing on drivers and actions that affect preparedness to climate change in the short-term works well for systems where there is a good understanding of the prominent drivers and potential impacts, such as flooding. This approach will, however, be more challenging to apply in the context of more complex systems, such as the natural environment, where the relationships are poorly understood and data may be limited. Over time the relevant drivers and actions may also change. To be aware of these changes, key stakeholders are regularly consulted to ensure that the assessment accurately reflects current understanding of the prominent drivers affecting UK's climate preparedness. Additional country examples to indicator development are summarised in Box 3.3.

Box 3.3. **Examples of national indicators used to monitor and evaluate adaptation**

Germany: The German monitoring framework focuses on climate change impacts and response indicators matched to the 15 action and cross-sectional fields prioritised in the German Adaptation Strategy (see the proposed indicator list in Chapter 7). The indicator system is based on six criteria (Schönthaler et al., 2010): i) it displays to the extent possible climatic impacts and adaptation, considers cause-effect-chains, and is accepted by experts; has a transparent prioritisation of the indicators given the complex and comprehensive nature of climate change; represents all 15 action and cross-sectional fields; ii) it can be implemented on the basis of existing data that will be collected in the future; has broad stakeholder engagement to facilitate the identification and application of a wide range of data; iii) it reflects available knowledge on the impacts of climate change and the effectiveness of adaptation measures by government departments as well as by non-governmental institutions and organisations; iv) it is open for regular review in response to evolving climate change knowledge and emerging political priorities; v) it links up with other indicator systems; and vi) it facilitates linkages with monitoring and reporting at the EU and the Länder level.

Australia: The assessment framework proposed for Australia's Climate Adaptation Outlook is based on the premise that decisions made today will determine the country's success in adapting to future climate change (Australian Government, 2013). It is therefore important that the risks are well understood, and that the governance structure (e.g. building codes, land-use planning and regulation of energy infrastructure) and market mechanisms (e.g. price signals and disclosure of climate risks) facilitate effective adaptation to both climate variability and change. Broad public acceptance is also a pre-requisite for action on climate change adaptation. To access progress, 12 indicators have been proposed (see Chapter 8).

Box 3.3. **Examples of national indicators used to monitor and evaluate adaptation** *(cont.)*

France: The French framework facilitates annual monitoring of progress made in achieving set objectives identified for 19 areas and one cross-sectoral theme outlined in the National Adaptation Plan (2011-15) (French Government, 2011). For each area and theme, an action sheet outlines five or six actions, each comprising several components that must be undertaken in that area, totaling 84 actions and 230 measures (see Chapter 9 for the complete set of indicators). These actions can be broadly categorised as i) production and dissemination of information, ii) adjustment of standards and regulations, iii) institutional adaptation, and iv) direct investment.

The indicator sets used to monitor and evaluate adaptation often have not been specifically designed for adaptation. Instead, they bring together a number of indicators that are already monitored on a regular basis and that together provide a good understanding of changes in the country's vulnerability to climate change within the context of national development objectives. This may entail a mix of qualitative outcome indicators and quantitative process indicators. On their own, any category of indicator may not be enough. For instance, a process indicator specifying whether a policy framework has been developed does not shed light on whether the policy has been implemented and what the corresponding outcomes are. It is useful to complement this type of indicator with qualitative indicators to assess how the policy may have contributed to changes observed (Lamhauge, Lanzi and Agrawala, 2012). Table 3.3 summarises the types of indicators currently being considered in eight developed and developing countries.

Table 3.3. **Types of indicators used to monitor and evaluate adaptation**

	Indicator categories				
	Climate change impacts	Exposure	Vulnerability	Adaptation process	Adaptation outcomes
Australia				√	√
France				√	√
Germany	√		√	*"Responses"*	*"Responses"*
Kenya				√	*"Vulnerability"* *"Adaptive capacity"*
Morocco	√		√	*"Adaptation"*	*"Adaptation"*
Nepal					√
Philippines				√	√
UK	√	*"Risk factors"*	*"Adaptation action"*		

Source: Adapted from GIZ (2013a), *Monitoring and Evaluating Adaptation at Aggregated Levels: A Comparative Analysis of Ten Systems*, Deutsche Gesellschaft für Internationale Zusammenarbeit, Eschborn.

To guide the development of indicators for adaptation, the standards developed for the United Nations Programme on HIV/AIDS can provide a useful starting point. These standards are intended to determine the quality and utility of proposed indicators, and to ensure that the indicators produce relevant information that can be used to inform national policy approaches (UNAIDS, 2010). Each of the six standards is complemented by a set of questions or key criteria to be considered when developing an indicator set (see Table 3.4).

Table 3.4. **Questions and information requirements to meet indicator standards**

STANDARD 1: The indicator is needed and useful	
Q. 1	Is there evidence that this indicator is needed at the appropriate level?
Q. 2	Which stakeholders need and would use the information collected by this indicator?
Q. 3	How would information from this indicator be used?
Q. 4	What effect would this information have on planning and decision-making?
Q. 5	Is this information available from other indicators and/or other sources?
Q. 6	Is this indicator harmonised with other indicators?
STANDARD 2: The indicator has technical merit	
Q. 1	Does the indicator have substantive merit?
Q. 2	Is the indicator reliable and valid?
Q. 3	Has the indicator been peer reviewed?
STANDARD 3: The indicator is fully defined	
Required information:	Title and definition Purpose and rationale Method of measurement Data collection methodology Data collection frequency Data disaggregation Guidelines to interpret and use data Strengths and weaknesses Challenges Relevant sources of additional information
STANDARD 4: Is it feasible to measure the indicator	
Q. 1	How well are the systems, tools and mechanisms that are required to collect, interpret and use data for this indicator functioning?
Q. 2	How would this indicator be integrated into a national monitoring and evaluation framework and system?
Q. 3	To what extent are the financial and human resources needed to measure this indicator available?
Q. 4	What evidence exists that measuring this indicator is worth the cost?
STANDARD 5: The indicator has been field-tested or used operationally	
Q. 1	To what extent has the indicator been field-tested or used operationally?
Q. 2	Is this indicator part of a system to review its performance in ongoing use?
STANDARD 6: The indicator set is coherent and balanced	
Q. 1	Does the indicator set give an overall picture of the adequacy or otherwise of the response being measured?
Q. 2	Does the indicator set have an appropriate balance of indicators across elements of the response?
Q. 3	Does the indicator set cover different monitoring and evaluation levels appropriately?
Q. 4	Does the set contain an appropriate number of indicators?

Source: UNAIDS (2010), *An Introduction to Indicators: UNAIDS Monitoring and Evaluation Fundamentals*, United Nations Programme on HIV/AIDS, Geneva.

The potential role of development co-operation providers

Development co-operation providers play an important role in assisting partner countries in formulating a national framework that is aligned with domestic adaptation priorities and their information needs. At the same time, partner countries often contribute some information at the local, regional or national level to the monitoring and evaluation of initiatives financed, at least in part, by development co-operation providers. These frameworks normally focus on the project and programme level rather than the national level. As a result, their emphasis differs from that observed in domestic frameworks for adaptation. For example, compared to the common focus of national indicators on climate change impacts and processes (see Table 3.3), one of the Key Performance Indicators for the UK International Climate Fund (ICF) aims to assess the proportion of people that have become more climate resilient as a result of support from the ICF. Similarly, the PPCR measures the extent to which climate change has been integrated into national and sector planning processes and the evidence available of strengthened government capacity and co-ordination mechanism to mainstream climate resilience.

The indicators collected by development co-operation providers contribute to meeting the reporting requirements they themselves are subjected to. They are, however, not always tailored to the partner country's policy-making cycle and as a result, may not reflect the country's adaptation priorities and information needs. This can be because domestic results frameworks may not be in place when financial support for adaptation is initiated. However, to reduce the reporting requirements faced by partner countries during this initial phase, co-operation providers should, to the extent possible, aim to align their results frameworks. This would be particularly beneficial for countries that receive support from multiple sources. A more co-ordinated approach would free up scarce domestic resources that instead could be allocated to the formulation and implementation of a domestic monitoring and evaluation framework.

Development co-operation providers can also play an important role in facilitating peer learning and enhancing partner countries' capacities to develop indicator sets to monitor adaptation. For example, through workshops and webinars relevant officials can discuss good practice approaches and share experiences. Alternatively, monitoring and evaluation officials in partner countries can spend some time with counterparts in developed countries to learn what approaches they have taken when developing and implementing an indicator set for adaptation. Further, to shed light on what approaches have already been tried when developing an indicator set for adaptation, development co-operation providers can put together a menu of indicators from different country contexts that partner countries may wish to consider (GIZ, 2014b).

Learning from adaptation approaches

It is only recently that countries have started explicitly focusing on adaptation within their national planning processes. National approaches to adaptation will be more effective if they are informed by the experience gained from existing efforts to manage climate change risks. Evaluations can provide a good understanding of how effective programmes, plans or policies are in achieving set objectives and generating lessons learned that can contribute to evidence-based policy processes. Evaluations can also guide decision makers on how to allocate scarce resources to activities known to deliver. The objectives of evaluations include (OECD, 2009):

- Demonstrating that policy aims are being achieved;
- Demonstrating that this is being done effectively and efficiently;
- Capturing lessons that can be learned to improve future delivery and decision making.

In the context of adaptation, where evaluations of national adaptation strategies or plans may not be feasible in the near future given the long time-horizons of climate change (Dinshaw et al., 2014), project or programme evaluations can inform adaptation planning and implementation processes. Although the challenges and uncertainties are similar at the national, project and programme levels, the more limited scope of individual projects and programmes can help to identify what approaches to adaptation are effective in enhancing climate resilience, and to better understand what the conditions required for their success may be. Evaluations of adaptation initiatives can be focused on particularly large adaptation interventions, at interventions that pilot particularly innovative approaches to adaptation, or a combination of the two. Box 3.4 outlines a number of questions that can guide governments when deciding which adaptation initiatives to evaluate.

Box 3.4. **Guiding questions to determine which adaptation initiatives to evaluate**

Four questions can provide guidance to national governments when deciding which programmes to evaluate:

- Is the programme of strategic relevance for national climate change adaptation? Interventions considered to be of particular importance in addressing climate change risks may be evaluated to ensure that this is indeed the case and to facilitate any adjustments in subsequent interventions if needed;
- Is the intervention testing an innovative approach to adaptation? Evaluation of pilot initiatives can help determine if they should be scaled-up;
- Is there evidence that a particular approach to adaptation is effective in reducing climate vulnerability or enhancing climate resilience and is it appropriate in different contexts? If this is not the case, an evaluation can provide valuable information as to how the intervention ought to be adjusted to suit different contexts;
- What impacts can be evaluated in the short- and medium-term? Will some outcomes only become apparent in the long-term? If so, should the final evaluation be delayed until the intervention is likely to show an effect? Alternatively, can proxy indicators that are linked to the planned outcomes but likely to show and effect earlier be used?

Source: Adapted from Khandker, S.R., G.B. Koolwal and H.A. Samad (2010), *Handbook on Impact Evaluation: Quantitative Methods and Practices*, World Bank, Washington, DC.

In LDCs, such project or programme evaluations could, for example, be linked to the adaptation priorities outlined in their NAPAs. These priorities usually constitute locally or regionally confined activities from which lessons can be derived. For example, financial support from the PPCR, the Adaptation Fund, or the Global Environment Facility is usually complemented by results frameworks tailored to the particular project or programme. The evaluations of these initiatives can shed light on what approaches to adaptation are effective in the given country context. Similarly, evaluations of other national initiatives that may not be labelled as adaptation or have primary objectives other than adaptation, but nonetheless contribute to reduced climate vulnerability, can also generate useful lessons.

Clear targets and objectives are important if projects and programmes are to facilitate learning. Furthermore, it is important that the evaluation framework is designed and implemented at the outset to set the stage for future evaluations. In line with good practice principles for evaluation of development assistance, the aim of the evaluation should be to determine the relevance and fulfilment of project or programme objectives, as well as the efficiency, effectiveness, impact and sustainability of the intervention (OECD, 1991). Box 3.5 provides some examples illustrating how the five principles might be applied to the context of adaptation.

To facilitate a process whereby evaluations inform adaptation planning processes, it is ideal if the incentives and capacities are in place to encourage producers and users of the

Box 3.5. **Key evaluative questions to include for adaptation interventions**

- **Relevance:** Does the policy or intervention address identified areas of likely vulnerability and risk? Are the assumptions or theory of change on which the activity is based logical or sensible in this context at this time? Are outputs consistent with the objectives of increasing resilience?
- **Efficiency:** Are activities cost efficient? Is this the most efficient way to improve adaptive capacity? Compare potential disaster costs vs. the cost of this particular approach to prevention (see e.g. GIZ, 2013b).
- **Impact:** What happened as a result of the adaptation policy? Why? What were the positive and negatives changes produced, directly or indirectly, intended or unintended? Did the intervention impact key areas of risk or affect resiliency factors?
- **Effectiveness:** To what extent were the objectives achieved? What factors contributed to achievements?
- **Sustainability:** Will benefits be maintained after the programme or support has ended? Do locals have ownership of the activity or programme, where possible? Have durable, long-term processes, structures and institutions for adaptation been created?

Source: Adapted from OECD (2008), *Evaluating Conflict and Peacebuilding Activities: Factsheet*, available at: *www.oecd.org/dac/evaluation/dcdndep/39289596.pdf.*

information to build on lessons learned in subsequent planning processes. Further, given the reliance of developing countries on external support for their adaptation planning and implementation, it is most useful if the findings from programme evaluations are available to national officials responsible for adaptation planning and budgeting. This, however, is difficult in practice, especially when institutional capacities are weak or when all the information is not digitised.

The potential role of development co-operation providers

Development co-operation providers have agreed on a number of aid effectiveness principles that, among others call for the use of partner countries' data and monitoring and evaluation systems. However, experience to date with country-led evaluations (CLEs) has been mixed. CLEs enable the partner country to own and lead the evaluation process by determining what policies or programmes will be evaluated, what questions will be asked, and how initiatives will be assessed (Segone, 2010). The mixed experience with CLEs suggests that development co-operation providers do not find the evaluations produced by partner countries' own systems sufficient for their accountability needs. This can either be in terms of inadequate quality of the evaluations, or because the CLEs are mainly focused on monitoring progress in implementing projects and programmes rather than assessing their effectiveness (Bedi et al., 2006).

The implementation of CLEs can also be hampered by weak institutional and human capacities in partner countries. The 2011 evaluation of the *Paris Declaration* found that capacity constraints were one of the main reasons why development providers continue to rely on their own monitoring and evaluation systems rather than using and strengthening partner countries own systems (Wood et al., 2011). Further complicating the matter is the fact that climate finance does not always go to a centralised government unit but is often fragmented with in-country responsibilities residing in different institutions (Miller, 2013). To overcome this challenge, some projects and programmes include funding earmarked for data collection to monitor and evaluate the initiative (Bedi et al., 2006). While this contributes to valuable learning, it may discourage government efforts to enhance domestic monitoring and evaluation capacity.

Alternatively, development co-operation providers may choose to work with senior government officials in partner countries to champion the importance of building domestic monitoring and evaluation capacity and to encourage the use of the information generated for national planning and budgeting processes. Further, development co-operation providers and partner countries can work together to jointly evaluate interventions. When this component is built in from the outset and collaboration continues beyond the end of the evaluation cycle to ensure that the findings feed into subsequent planning processes, it can build interest in and demand for better evidence about results. Over time, this will contribute to improved domestic evaluation capacity, and in turn, enhance the value of CLEs. In practice, however, a more common approach has been for development co-operation providers operating in the same area to jointly evaluate their activities.

National audits and climate expenditure reviews

An increasing priority for developing countries and for development co-operation providers is to understand if the government's adaptation policy meets international and national commitments, and if they are being met cost-effectively. This was, for example, the focus of the 2012 Global Forum on Development Effectiveness. The Forum explored how lessons learned from development effectiveness can be transferred to the context of climate finance, as resources are increasingly being earmarked for either mitigation or adaptation objectives under the UNFCCC (Global Forum, 2012). In this context, audits and Climate Public Expenditure and Institutional Reviews (CPEIRs) can play an important role in establishing the flow of financial resources for adaptation.

Supreme audit institutions (SAIs)[3] are responsible for ensuring that public funds are spent effectively and in compliance with existing rules, regulations and principles of good governance. In particular, SAIs have the responsibility to "provide legislatures and their citizens with the information they need to hold governments accountable for prudent financial management, and to varying degrees for compliance with domestic laws and international agreements, policy implementation, and programme performance" (INTOSAI, 2010a). Audits of climate change policies have in recent years become a priority of SAIs, in part due to the relatively recent focus on climate change, but also due to the challenges SAIs face when auditing adaptation and mitigation policies (see Box 3.6).

Box 3.6. **Global and regional audits on government responses to climate change**

The International Organisation for Supreme Audit Institutions (INTOSAI) brings together supreme audit institutions (SAIs) in United Nations member countries or its specialised agencies. INTOSAI has a number of thematic working groups, including the Working Group on Environmental Auditing (WGEA), responsible for auditing climate change. INTOSAI WGEA published in 2010 a comparative study that examined different national approaches to auditing climate change programmes and performance in 14 member countries.[1] The SAI in each country carried out domestic audits in response to the country's climate change priorities and national standards and regulations. The European Organisation of Supreme Audit Institutions (EUROSAI) – one of seven regional working groups of the INTOSAI –

Box 3.6. **Global and regional audits on government responses to climate change** *(cont.)*

undertook a similar comparative audit in collaboration with SAIs in nine European countries[2] that assessed their governments' preparedness to climate change and actions taken to adapt to it.

The audits covered a variety of topics ranging from national compliance with international commitments on climate change, the assessment of risks and vulnerabilities, the co-ordination and management arrangements in place, the availability of reliable information, and the performance of the policy instruments used. The audits identified the strengths and weaknesses in governments' responses to climate change that in turn contributed to some governments introducing changes to their national approaches. The audits demonstrated that robust climate change risk and vulnerability assessments are still at a relatively early stage and that national initiatives primarily focus on current climate variability rather than projected climate change. The audits also noted that weak management structures adversely affect the co-ordination and alignment of adaptation initiatives across sectors and levels of government. Finally, robust climate change data is often lacking, preventing the government from making informed decisions on priority adaptation needs and to monitor progress over time.

1. Australia, Austria, Brazil, Canada, Estonia, Finland, Greece, Indonesia, Norway, Poland, Slovenia, South Africa, the UK, and the US.
2. Austria, Bulgaria, Cyprus, Malta, the Netherlands, Norway, Russia and Ukraine.
   Note by Turkey: "The information in this document with reference to 'Cyprus' relates to the southern part of the Island. There is no single authority representing both Turkish and Greek Cypriot people on the Island. Turkey recognises the Turkish Republic of Northern Cyprus (TRNC). Until a lasting and equitable solution is found within the context of the United Nations, Turkey shall preserve its position concerning the 'Cyprus issue'".
   Note by all the European Union Member States of the OECD and the European Union: "The Republic of Cyprus is recognised by all members of the United Nations with the exception of Turkey. The information in this document relates to the area under the effective control of the Government of the Republic of Cyprus".

Source: INTOSAI (2010a), *Coordinated International Audit on Climate Change: Key Implications for Governments and their Auditors*, International Organisation of Supreme Audit Institutions, Working Group on Environmental Auditing; EUROSAI (2012), *Adaptation to climate change – are governments prepared? A cooperative audit*, European Organisation of Supreme Audit Institutions Working Group on Environmental Auditing, Oslo.

The objectives of national climate change audits (here focusing on the adaptation components) have been grouped into three categories (INTOSAI, 2010b):

- **International agreements:** Does the government's response to adaptation meet international agreements? The UNFCCC and the Kyoto Protocol include a number of commitments that signatories sign up to. For example, the Convention states that "all Parties [shall] formulate, implement, publish and regularly update national and, where appropriate, regional programmes containing measures to [...] facilitate adequate adaptation to climate change" (UNFCCC, 1992, Article 4, paragraph 1b). Further "all Parties [shall] co-operate in preparing for adaptation to the impacts of climate change" (UNFCCC, 1992, Article 4, paragraph 1e).
- **Good governance:** Is the government's response to adaptation co-ordinated and based on clearly defined roles and responsibilities? Audits can also examine the transparency of decision-making processes, the level of engagement of stakeholders, and where appropriate, the extent to which adaptation initiatives are managed by objectives and results.
- **Good management:** How good is the management of the government's response to adaptation? Are the organisational structures, authorities, and human resources

suitable for managing the risks? Are the objectives and targets well defined and prioritised and do they reflect projected risks? Are activities established to address identified risks and to achieve set objectives? Is the management of established risks well communicated and, when necessary, revised? And are monitoring mechanisms in place that provide regular information of progress made?

The audits may include an examination of whether costs and benefits have been estimated, the extent to which programmes, plans or strategies address both short-term variability and long-term change, whether expected results are being achieved, and if the government is on track to meet national and international commitments. Furthermore, national adaptation audits can examine if national authorities are adequately monitoring and evaluating performance, if findings are reported in a transparent manner, and if financial resources are properly administered and reach the intended recipients. Table 3.5 outlines a number of questions auditors may consider when auditing adaptation. Box 3.7 summarises the findings from three Brazilian audits on the government's approach to adaptation. The Box highlights the kind of information governments can derive from such audits and some of the challenges countries face when adapting to climate change.

Table 3.5. **Key questions for adaptation audits**

STEP 1: Get an overview of the country's vulnerability to climate change

- What are the actual and potential impacts of climate change (on a national, sectoral or thematic scale)?
- What is the adaptive capacity (i.e. the ability of a system to respond successfully to climate variability and change)?
- What is the vulnerability to climate change (determined by the country's actual and potential impacts and adaptive capacity)?
- Are the risk and vulnerability assessments of sufficient quality (including uncertainty estimates, financial estimates)?

STEP 2: Map the government response in adapting to climate change

- What are the objectives and targets of adaptation policies (drawing on national and international commitments on adaptation but also in the context of sustainable development and other multilateral environmental agreements)?
- What are the policy instruments in place that directly or indirectly address the country's adaptation priorities and to what extent can they be used for compliance or performance audits?
- Is the policy framework of sufficient quality (based on risk and vulnerability assessments, including targets and responsibilities, timeframes and a monitoring and evaluation framework)?
- Who are the public players and what are their roles and responsibilities?
- Has a co-ordinating body been established and do ministries, agencies and other stakeholder perform their tasks in accordance with established roles and responsibilities?

STEP 3: Choose audit topics and priorities

- Has the government assessed the key vulnerabilities in a proper manner (and how reliable is the data used)?
- Has the government developed a plan or strategy that adequately addresses the climate risks identified in Step 1 and 2?
- Has the government addressed the need for climate change action in the most vulnerable sectors and areas?
- Are the adaptation measures socially, economically and/or environmentally sustainable?
- Are the financial resources misstated? Is the budget spent as intended?
- Are appropriate actions being carried out to adapt to the identified vulnerabilities (are objectives, roles and responsibilities clear, does the government have a strategy in place to address identified barriers)?
- Is the government focusing on keeping the costs of adaptation as low as possible?

STEP 4: Design the audit

- Have the responsible ministries identified climate change-related threats (have risk assessments been conducted and have these been subjected to quality control, review and a consultation process)?
- Does the government have in place an overarching policy, plan or strategy that responds to identified climate change risks?
- Is the adaptation governance efficient?
- Are policy instruments effective?

Source: Adapted from INTOSAI (2010b), *Auditing the Government Response to Climate Change*, International Organisation of Supreme Audit Institutions, Working Group on Environmental Auditing; EUROSAI (2012), *Adaptation to climate change – are governments prepared? A cooperative audit*, European Organisation of Supreme Audit Institutions Working Group on Environmental Auditing, Oslo.

Box 3.7. **Brazilian audit of adaptation in different sectors**

Brazil has undertaken three audits on the government's approach to adaptation in areas at particular risk from climate change: i) in the livestock and agricultural sector; ii) in coastal zones, and iii) in water security in Brazil's semi-arid region. Each audit covered the period up to the end of 2008 and examined the main vulnerabilities in the specific areas, the potential risks from climate change under different climate scenarios, and the extent to which identified risks where matched by corresponding government initiatives consistent with good practice approaches on co-ordination, integration, governance and accountability.

In the agricultural sector, the audit concluded that potential climate change risks were not properly identified due to inadequate access to meteorological data. A large proportion of meteorological data is recorded on paper, making it inaccessible, and in turn affecting the quality of climate models. Further, given the relatively early stage of adaptation in the agricultural sector, clear guidelines on how agencies should integrate adaptation into their planning and implementation processes are not yet in place. Finally, the sector did not meet expected standards on co-ordination, integration, governance and accountability. In particular the lack of clear allocation of roles and responsibilities between public agencies and institutions was raised. To address these issues, the audit recommended that the National Plan on Climate Change introduces guidelines that clearly specify sectoral adaptation priorities, that the meteorological data becomes digitised and easily accessible, and that clear instructions to public managers are developed that explain how climate change scenarios can be used to inform the planning and implementation of agricultural policies.

Similarly, the climate change risks and vulnerabilities are not adequately understood in the context of coastal zones. The monitoring and storage of data on oceanic variables is currently decentralised with monitoring carried out by a number of public institutions, universities and research institutes. As a result, some oceanic variables crucial for constructing robust scenarios are not monitored. Further, the National Plan on Climate Change does not provide specific guidelines on adaptation in coastal zones. Public policies in relevant sectors (e.g. marine shipping and civil defense) are just starting to address the issue. The audit recommended that an action plan for monitoring oceanic variables gets developed, that a data bank to store the information is established, and that the National Council of Water Resources and the National Council of Environment take relevant actions in their respective areas.

On water security in Brazil's semi-arid region, the audit found that there is no national risk assessment available, that water management and distribution policies do not take the potential impacts of climate change into account, and that roles and responsibilities are not clearly defined. To overcome these challenges, the audit recommended that the institutions responsible for implementing the National Plan on Climate Change promote good co-ordination between relevant institutions and across sectors to produce a risk assessment and to encourage technical research on the potential impacts of climate change on water resources in Brazil's semi-arid region. Further, it recommended establishing an alert system for drought and desertification, to develop a regional climate change scenario, and for the responsible institutions to use this information to plan and implement climate resilient water resources policies.

Source: INTOSAI (2010a), *Coordinated International Audit on Climate Change: Key Implications for Governments and their Auditors*, International Organisation of Supreme Audit Institutions, Working Group on Environmental Auditing.

Complementing national audits on adaptation, CPEIRs were introduced in 2011. CPEIRs assist developing countries in reviewing their policy response to climate change, evaluating if the institutional mechanisms in place are effective in delivering climate finance, and assessing if public expenditures are aligned with identified objectives. CPEIRs are based on broad stakeholder consultation that extends beyond the environment agencies to also engage central planning and finance agencies in the discussion of national climate change policies and their financial implications (Bird et al., 2012). This contributes to an institutional and policy environment that is well informed of the climate change challenges and is in a strong position to respond to identified challenges through the integration of climate change considerations into national planning and budgeting process (Aid Effectiveness, n.a.).

The CPEIR includes an assessment of fiscal sustainability, resource allocation, the role of government, the efficiency and effectiveness of spending, institutional capacity, and the alignment of incentives (Aid Effectiveness, n.a.). Some key questions to consider when doing a CPEIR analysis are summarised in Table 3.6. There is some overlap between these

Table 3.6. **Questions to consider for a Climate Public Expenditure and Investments Review**

KEY QUESTIONS for CPEIR policy analysis

- What level of engagement does the country have with the international policy discourse within the UNFCCC?
- How much policy attention does climate change receive within national development planning?
- Are there explicit funding strategies for climate change actions (e.g. in costed action plans)?
- What is the overall coherence of the national response to climate change across a range of sectors?
- Does climate change appear as an emerging policy theme in cross cutting government programmes (e.g. social protection/livelihoods/agriculture/infrastructure, etc.)?
- Is climate change a policy theme at the local government level?
- Does climate change policy recognise the role of communities, the private sector, civil society and the media in ensuring multi-stakeholder participation in climate change initiatives?
- Is there a monitoring and evaluation system for climate change actions that goes beyond the measurement of financial inputs?

KEY QUESTIONS for CPEIR institutional analysis

- Is there clarity over the roles and responsibilities for climate change between different government departments within and between ministries?
- Have new organisations been created to address climate change issues and, if so, how do such structures interact with existing government ministries, departments and agencies?
- Are the organisational structures compatible with these policy and strategy objectives as well as their legal mandates? How formalised are these structures?
- Does institutional collaboration and coordination on climate change need to be strengthened? And, if so, how can it be done?
- What is the level of engagement of the national legislature? What role does parliament play (through specialist committees) in overseeing the government's climate change programmes?
- What is the capacity of local government to fulfil any service delivery role?

KEY QUESTIONS for CPEIR expenditure analysis

- What are the characteristics of the national public finance management system within which spending on climate change-related actions occur?
- What is the state of the government's overall financial position: is there "fiscal space" to support the allocation of resources towards climate change actions?
- What are the trends in public expenditure generally and specifically for climate change actions?
- Where is climate change related expenditure happening across government ministries/departments/agencies?
- What level of expenditure has as its primary objective the delivery of specific outcomes that improve climate resilience or contribute to mitigation actions?
- What is the level of climate change-related expenditure across any economic and functional classifications of the budget?
- What is the level of public expenditure on climate change actions at the local government level?
- What are the main sources of funding for climate change actions? What role do extra-budgetary funds play? What role do international sources of climate finance play?

Table 3.6. **Questions to consider for a Climate Public Expenditure and Investments Review** *(cont.)*

KEY QUESTIONS for the CPEIR sub-national analysis

- What is local government's understanding of, and contribution to, addressing climate change?
- What are the main sources of funding for local level climate change-related actions?
- What is local government's capacity to prioritise, manage and deliver climate finance based on national and local climate change priorities and institutional arrangements?
- What other local stakeholders are involved in the delivery of climate finance?
- What accountability framework exists for delivering climate finance at the local level?

Source: Bird, N. et al. (2012), "The Climate Public Expenditure and Institutional Review (CPEIR): A methodology to review climate policy, institutions and expenditure'", *UNDP/ODI Working Paper.*

and the questions summarised above for adaptation audits. The main difference is the explicit focus on the country's institutional framework for climate change finance and the emphasis on an expenditure analysis in addition to the policy assessment. However, unlike national audits, governments are not legally bound to respond to CPEIRs findings. The processes can therefore complement each other.

Since the CPEIR methodology was first piloted in Nepal in 2011 (Government of Nepal, 2011), reviews have been undertaken in Bangladesh, Thailand, Samoa and Cambodia. Reviews are currently underway in Timor-Leste and Vietnam while others have been planned in Latin America, the Caribbean region and in Africa. The CPEIRs have served as building blocks for the development of climate fiscal frameworks that assess the demand and supply of climate finance reflecting both domestic and external sources. Further, CPEIR can help governments to improve the prioritisation, efficiency and effectiveness of public resources allocated for climate change actions (Government of Nepal, 2011).

The potential role of development co-operation providers

In an effort to enhance the capacity of SAIs in partner countries, 15 development co-operation providers[4] signed in 2009 a Memorandum of Understanding with INTOSAI. The memorandum calls for a more strategic and co-ordinated approach to support provided to SAIs. Specifically, it states that the 15 development providers will support the "strengthening of public financial management in partner countries, including the external governmental auditing function, with a view to ensuring that public resources are properly used and that funding reaches the intended end user" (INTOSAI, 2009, 4). To achieve these objectives, the SAIs in partner countries agree to develop individual country-led strategic action plans. Providers of development support, in turn, commit to respect the leadership of partner countries' SAIs, their independence and autonomy in developing and implementing their strategic action plans, and to mobilise additional funds to strengthen the capacity of SAIs through better and more effective support initiatives.

Capacity building initiatives can take many forms, including exchanges with partner SAIs, workshops and peer reviews (INTOSAI, 2007). The following measures can assist staff in development co-operation agencies in better understanding the role of SAIs and how they can support them to play a more effective oversight role (OECD, 2011, 9):

- Develop and support long-term capacity development projects for SAIs based on detailed assessments of their political context and strategic plans. In the context of adaptation, such support will be particularly important since many developing countries have or are in the process of developing national adaptation plans or strategies, but implementation is still at an early stage;

- Engage SAIs in auditing projects supported by providers of development co-operation – providing coaching and training support where needed. In the context of adaptation, this will entail close collaboration to better understand what the priority adaptation needs are;
- Advocate on behalf of SAIs with developing country governments, parliaments, civil society organisations and others, helping raise the profile of SAIs and encourage the use of audit findings. In the context of adaptation, where a specific focus on adaptation is still relatively recent, continuous learning, and the use of lessons learned for national planning and budgeting processes is crucial;
- Use the results of SAI audits in budget negotiations to ensure that national audits contribute to positive change. In the context of adaptation, this can also help to ensure that budget allocations for adaptation are channelled to evidence-based policy processes.

Development co-operation providers may consider investing resources to establish and strengthen links with partner country SAIs, and their stakeholders, to ensure that their support is aligned with domestic needs. In doing so, it can be helpful to collaborate with their domestic SAI counterparts or other experts (OECD, 2011).

Notes

1. The countries included in the international review were: Australia, Austria, Brazil, Canada, Estonia, Finland, Greece, Indonesia, Norway, Poland, Slovenia, South Africa, the United Kingdom, and the United States (INTOSAI, 2010a). The countries in the European review were: Austria, Bulgaria, Cyprus (for notes on Cyprus, see notes in Box 3.7), Malta, the Netherlands, Norway, Russia and Ukraine (EUROSAI, 2012).
2. The seven strategies priorities are: i) food security, ii) water sufficiency, iii) ecological and environmental stability, iv) human security, v) climate smart industries and services, vi) sustainable energy, and vii) knowledge and capacity development.
3. Alternatively referred to as National Audit Office, Court of Audit, Audit Board, or Office of the Auditor General.
4. This has subsequently increased to 20 providers of development support and includes: African Development Bank, Asian Development Bank, Australian Agency for International Development, Austrian Development Agency, Belgian Ministry of Foreign Affairs, Canada, European Commission, GAVI Alliance, Global Fund to Fight AIDS, Tuberculosis and Malaria, Inter-American Development Bank, International Monetary Fund, Ireland, Islamic Development Bank, Netherlands Ministry of Foreign Affairs, Norwegian Agency for Development Cooperation, Switzerland, Sweden, United Kingdom, United States of America, and the World Bank.

References

Aid Effectiveness (n.a.), *Climate Change Public Expenditure Review Sourcebook*, available at: *www.aideffectiveness.org/CPEIR*.

Australian Government (2013), *Climate Adaptation Outlook: A Proposed National Adaptation Assessment Framework*, Department of Industry, Innovation, Climate Change, Science, Research and Tertiary Education, Commonwealth of Australia, Canberra.

Australian Government (2006), *Climate Change Impacts and Risk Management: A Guide for Business and Government*, Australian Greenhouse Office, Department of the Environment and Heritage, Canberra.

Bedi, T. et al. (2006), *Beyond the Numbers: Understanding the Institutions for Monitoring Poverty Reduction Strategies*, World Bank, Washington, DC.

Bird, N. et al. (2012), "The Climate Public Expenditure and Institutional Review (CPEIR): A methodology to review climate policy, institutions and expenditure", *UNDP/ODI Working Paper*.

Cardona et al. (2012), "Determinants of risk: exposure and vulnerability", in *Managing the Risks of Extreme Events and Disasters to Advance Climate Change Adaptation*, Field et al. (eds.), a Special Report of Working Groups I and II of the Intergovernmental Panel on Climate Change (IPCC), Cambridge University Press, Cambridge and New York.

Casado-Asensio, J. and R. Steurer (2013), "Integrated strategies on sustainable development, climate change mitigation and adaptation in Western Europe: Communication rather than co-ordination", *Journal of Public Policy*, *http://dx.doi.org/10.1017/S0143814X13000287*.

Defra (UK Department for Environment, Food and Rural Affairs) (2012), *The UK Climate Change Risk Assessment 2012: Evidence Report*, Defra, London.

Dinshaw, A. et al. (2014), "Monitoring and Evaluation of Adaptation: Methodological Approaches", *OECD Environment Working Papers*, No. 74, OECD Publishing, Paris, *http://dx.doi.org/10.1787/5jxrclr0ntjd-en*.

EUROSAI (European Organisation of Supreme Audit Institutions) (2012), *Adaptation to climate change – are governments prepared? A cooperative audit*, EUROSAI Working Group on Environmental Auditing, Oslo.

French Government (2011), *National Plan Climate Change Adaptation*, Ministry of Ecology, Sustainable Development, Transport and Housing, Paris.

GIZ (Deutsche Gesellschaft für Internationale Zusammenarbeit) (2014a), *The Vulnerability Sourcebook – Concept and Guidelines for Standardized Vulnerability Assessments*, GIZ, Eschborn.

GIZ (2014b), *Repository of national-level adaptation M&E indicators*, GIZ, Eschborn.

GIZ (2013a), *Monitoring and Evaluating Adaptation at Aggregated Levels: A Comparative Analysis of Ten Systems*, GIZ, Eschborn.

GIZ (2013b), *Saved health, saved wealth: an approach to quantifying the benefits of climate change adaptation*, GIZ, Eschborn.

GIZ (2013c), *Comparative analysis of climate change vulnerability assessments: Lessons from Tunisia and Indonesia*, GIZ, Eschborn.

Global Forum (2012), "Climate Public Expenditure and Institutional Reviews (CPEIRs): Past Experience and the Way Forward", a Global Forum facilitated by the Partnership for Action on Climate Change Finance and Effective Development Co-operation, held on 10-12 September 2012, Bangkok, Thailand.

Government of Nepal (2011), *Nepal Climate Public Expenditure and Institutional Review (CPEIR)*, National Planning Commission, United Nations Development Programme, United Nations Environment Programme, Capacity Development for development Effectiveness Facility for Asia Pacific, Kathmandu.

Hammill, A. and T. Tanner (2011), "Harmonising Climate Risk Management: Adaptation Screening and Assessment Tools for Development Co-operation", *OECD Environment Working Papers*, No. 36, OECD Publishing, Paris, *http://dx.doi.org/10.1787/5kg706918zvl-en*.

Harvey, A. et al. (2011), *Provision of research to identify indicators for the Adaptation Sub-Committee*, AEA, Edinburg.

IIED (International Institute for Environment and Development) (2013), *Tracking Adaptation and Measuring Development (TAMD) in Ghana, Kenya, Mozambique, Nepal and Pakistan: Meta-analysis findings from Appraisal and Design phase*, IIED, London.

INTOSAI (International Organisation of Supreme Audit Institutions) (2010a), *Coordinated International Audit on Climate Change: Key Implications for Governments and their Auditors*, INTOSAI, Working Group on Environmental Auditing.

INTOSAI (2010b), *Auditing the Government Response to Climate Change*, INTOSAI, Working Group on Environmental Auditing.

INTOSAI (2009), *Memorandum of Understanding between the International Organisation of Supreme Audit Institutions (INTOSAI) and the Donor Community*, available at: *www.idi.no/artikkel.aspx?MId1=15&AId=497*.

INTOSAI (2007), *Building capacity in Supreme Audit Institutions: A guide*, INTOSAI, Capacity Building Sub-Committee 1.

Khandker, S.R., G.B. Koolwal and H.A. Samad (2010), *Handbook on Impact Evaluation: Quantitative Methods and Practices*, World Bank, Washington, DC.

Lamhauge, N., E. Lanzi and S. Agrawala (2012), "Monitoring and Evaluation for Adaptation: Lessons from Development Co-operation Agencies", *OECD Environment Working Papers*, No. 38, OECD Publishing, Paris, *http://dx.doi.org/10.1787/5kg20mj6c2bw-en*.

LEG (Least Developed Countries Expert Group) (2012), *National Adaptation Plans. Technical guidelines for the national adaptation plan process*, LEG, UNFCCC Secretariat, Bonn.

LEG (2002), *Annotated guidelines for the preparation of national adaptation programmes of action*, LEG, UNFCCC Secretariat, Bonn.

Miller, M. (2013), *Climate Public Expenditure and Institutional Reviews (CPEIRs) in the Asia-Pacific Region. What Have We Learnt?*, Capacity Development for Development Effectiveness Facility and United Nations Development Programme.

OECD (Organisation for Economic Co-operation and Development) (2013), *Water Security for Better Lives*, OECD Studies on Water, OECD Publishing, Paris, *http://dx.doi.org/10.1787/9789264202405-en*.

OECD (2011), *Good Practices in Supporting Supreme Audit Institutions*, OECD Publishing, Paris.

OECD (2009), *Implementation Guidelines on Evaluation and Capacity Building for the Local and Micro Regional Level in Hungary: A Guide*, available at: *www.oecd.org/cfe/leed/42748793.pdf*.

OECD (2008), *Evaluating Conflict and Peacebuilding Activities: Factsheet*, available at: *www.oecd.org/dac/evaluation/dcdndep/39289596.pdf*.

OECD (1991), *Principles for Evaluation of Development Assistance*, OECD Publishing, Paris.

Republic of Kenya (2012), *National Performance and Benefit Measurement Framework: Section A: NPBMF and MRV+ System Design, Roadmap and Guidance*, Ministry of Environment and Mineral Resources, Kenya.

Schönthaler, K. et al. (2010), *Establishment of an Indicator Concept for the German Strategy on Adaptation to Climate Change*, available at: *www.uba.de/uba-info-medien-e/4031.html*.

Segone, M. (2008), "Evidence-based policy making and the role of monitoring and evaluation within the new aid environment", in M. Segone (ed.), *Bridging the gap: The role of monitoring and evaluation in evidence-based policy making*, UNICEF, the World Bank, and the International Development Evaluation Association, Geneva.

UNAIDS (United Nations Programme on HIV/AIDS) (2010), *An Introduction to Indicators: UNAIDS Monitoring and Evaluation Fundamentals*, UNAIDS, Geneva.

UNFCCC (United Nations Framework Convention on Climate Change) (2000), *UNFCCC guidelines on reporting and review*, Report of the United Nations Framework Convention on Climate Change Conference of the Parties on its fifth session, held in Bonn 25 October-5 November, available at: *https://unfccc.int/files/national_reports/annex_i_natcom/_guidelines_for_ai_nat_comm/application/pdf/01_unfccc_reporting_guidelines_pg_80-100.pdf*.

UNFCCC (1992), *United Nationas Framework Convention on Climate Change*, United Nations, Bonn.

Wood, B. et al. (2011), *The Evaluation of the Paris Declaration, Final Report*, Danish Institute for International Studies, Copenhagen.

PART II

Emerging country indicators to monitor and evaluate adaptation

National Climate Change Adaptation
Emerging Practices in Monitoring and Evaluation

PART II

Chapter 4

Proposed indicators for Kenya's climate change action plan

Kenya's National Climate Change Action Plan (NCCAP) covers both mitigation and adaptation. A complementary National Performance and Benefits Measurement Framework (NPBMF) has been proposed. The objective of the framework is to track both mitigation and adaptation actions and the synergies between the two. It is informed by a methodology developed by the International Institute for Environment and Development (IIED) called Tracking Adaptation and Measuring Development (TAMD). The framework combines top-down indicators that assess institutional (adaptive) capacity and bottom-up indicators that measure vulnerability. The proposed indicators are linked to national level indicators already being measured on a regular basis.

Top-down institutional adaptive capacity indicators

The analysis undertaken for the preparation of the NCCAP proposed over 300 adaptation actions. To monitor these actions, 63 national level, process-based indicators measuring institutional adaptive capacity were identified. Based on this set of indicators, 28 county level, outcome-based indicators were identified. Through stakeholder consultation, this number was subsequently reduced to 10. The objective of these indicators is to measure the effectiveness of national initiatives to build institutional adaptive capacity at the county level. Although most of the actions in the NCCAP will take place at the national level, it is desirable to measure institutional adaptive capacity at the county level since that is where adaptive capacity translates into practical benefits for the people of Kenya. The 10 county level indicators and the complementary 63 process-level indicators are outlined below. A sample data sheet outlines how the county level indicators are measured in practice.

Proposed county level institutional adaptive capacity (top-down) indicators

Ref. No.	Proposed county level indicator
1	% of county roads that have been made "climate resilient" or that are not considered to be vulnerable [2, 3, 4, 5, 6]
2	% of new hydroelectric projects in the county that have been designed to cope with climate change risk [7, 8, 9, 10, 11]
3	% of population by gender and areas subject to flooding and/or drought in the county who have access to information from [Kenya Meteorological Department] on rainfall forecasts [12, 13, 14, 15, 16, 20]
4	% of people by gender in the county permanently displaced from their homes as a result of flood, drought or sea-level rise [21, 22, 23]
5	% of poor farmers and fishermen in the county with access to credit facilities or grants [31]
6	% of total livestock numbers killed by drought in the county [32, 33, 41]
7	% of area of natural terrestrial ecosystems in the county that have been disturbed or damaged [43, 44, 46]
8	% of water demand that is supplied in the county [23, 44, 48, 56]
9	% of poor people by gender in drought prone areas in the county with access to reliable and safe water supplies [23, 44, 50, 46]
10	Number of ministries at county level that have received training for relevant staff on the costs and benefits of adaptation, including valuation of ecosystem services [62, 63]

Note: The number in [square brackets] are the reference numbers for national level indicators (outlined in the table below) to which these county level indicators relate.

Source: Republic of Kenya (2012a), *National Performance and Benefit Measurement Framework: Section B: Selecting and Monitoring Adaptation Indicators*, Ministry of Environment and Mineral Resources, Kenya.

Proposed nation level, process-based indicators on institutional adaptive capacity

Ref. No.	Proposed national indicators (majority processed-based)
1	Number of new and existing port and harbour facilities designed to cope with rising sea levels
2	Km of the national existing and proposed new road network (including bridges and culverts) that has been assessed for vulnerability to flooding, river or coastal erosion or landslide
3	Climate change impacts relating to transport explicitly addressed in the next National Spatial Plan
4	Ksh/year required for increasing climate resilience of the road network allocated in the National Integrated Transport Master Plan (NITMP)
5	Number of transport authority staff attending training courses on road infrastructure design modifications to enhance climate resilience
6	Number of new projects that upgrade the road network specifically to increase resilience to flooding, erosion or landslides
7	Number of staff involved in hydroelectric asset design, identification of sites for generating capacity, identification of sites for substations, transmission lines or procurement of the above trained in assessing climate impacts & response strategies
8	Climate change impacts relating to critical energy infrastructure explicitly addressed in the next National Spatial Plan
9	Number of new hydroelectric power projects that have been assessed for vulnerability to drought/ low water levels under future climate scenarios; % of substations that have been assessed for vulnerability to flooding
10	% of water catchments serving hydropower facilities for which a climate sensitive management plan has been implemented
11	Ksh/year allocated to collaborative initiatives involving Ministries of Energy and Water at national level
12	Ksh/year allocated to dissemination of information on drought and rainfall to vulnerable communities
13	Number of new fully operational weather stations reporting accurate data to Kenyan Meteorological Department (KMD) or % increase in the country covered by the KMD observational network
14	Number of climate datasets available without costs to the general public through the KMD website or % number of technical ministries, NGOs, private sector stakeholders accessing data without costs
15	Ksh/year allocated to capacity building on climate modelling research and necessary IT assets (including funds from international sources) or evidence of regional climate model downscaling to national and county levels
16	Number of urban development plans that incorporate disaster risk reduction actions for poor communities, with specific recognition of the problems faced by women
17	Number of public buildings, emergency services and associated facilities screened for climate vulnerability with a flexible and costed response action plan
18	Climate resilience relating to building plans is reflected in the First National Spatial Plan
19	Number of current Information and Communication Technology (ICT) assets screened for climate vulnerability with a flexible and costed response action plan
20	Number of internet and mobile applications being used to access climate information
21	Ksh/year allocated by National Council for Science and Technology (NCST) to cross-sectoral research projects relating to climate change vulnerability or adaptation; % of above research project funding allocated to proactive dissemination of research results to poor communities
22	Ksh/year allocated to implementation of a national action plan addressing climate related migration
23	Number of transboundary agreements which integrate climate risk over water resources signed with neighbouring countries
24	Tourism sector climate change adaptation actions integrated into the First National Spatial Plan
25	National Tourism Policy reflects climate risk and vulnerability and encourages appropriate adaptation
26	% of the key national existing and proposed new tourist developments that has been assessed for vulnerability to flooding or drought
27	Ksh/year allocated to a research programme on climate change and tourism
28	Number of agriculture/ livestock/ fishery extension staff trained in geographically specific climate resilience strategies
29	Ksh/year allocated to market access improvement projects
30	Ksh/year allocated to rolling out additional crop, livestock and fishery insurance projects
31	Ksh/year allocated to supporting private sector loan facilities and grants to help poor farmers during climate induced hardship
32	Climate change adaptation reflected in the rangelands policy and action plan
33	Ksh/year allocated to the development of water resources that support climate change adaptation in the rangelands
34	Climate resilience reflected in the revised fisheries policy and relevant legislation
35	Ksh/year allocated to research programme on fisheries and climate change
36	Number of new marine protected areas gazetted
37	Number of business continuity insurance schemes covering extreme climate events available
38	Number/year of joint climate change meetings held between the Ministry of Public Health and Sanitation (MPHS) and the Ministry of Water and Irrigation MWI)
39	Ksh/year allocated to activities directed at controlling malaria in a changing climate
40	Number of climate and risk vulnerability assessments undertaken for various health subsectors
41	Ksh allocated to risk assessments for critical dry season areas and the development of adaptation actions for these areas
42	Net number/year of new climate resilient trees planted minus mortality from last year
43	Number of economic ecosystem valuations undertaken for critical ecosystems with recommendations on resilience building

Proposed nation level, process-based indicators on institutional adaptive capacity *(cont.)*

Ref. No.	Proposed national indicators (majority processed-based)
44	Number of institutional work plans (e.g. Kenyan Water Service and Kenyan Forest Service) that contain wildlife adaptation strategy actions
45	Ksh/year allocated for legal actions against illegal encroachment into protected areas
46	Fire management plans for protected and non-protected areas incorporate enhanced preparedness/actions for climate induced fires in terms of additional human, financial and technical resources
47	Ksh/year allocated for water storage capacity development, inter-basin transfers and exploitation of deep aquifers
48	% of water catchments for which demand management plans exist
49	Number of farmers receiving information on soil and water conservation and slope stabilisation
50	Number of water authorities that have integrated climate change impacts in their design, operation and maintenance of water assets
51	Number of sanitation authorities that have integrated climate change impacts in their design, operation and maintenance of sanitation assets
52	Ksh/year allocated to climate related DRR in water sector plans
53	Number of climate vulnerability and risk assessments undertaken in the water sector
54	% of water catchments with a climate change risk assessment incorporated in the catchment management plan
55	Ksh/year allocated to the implementation of water efficiency measures at a national level
56	Number of water stations for which data are officially reported and analysed
57	Number of new urban housing developments with flood mitigation measures in place
58	Number of current critical urban and housing infrastructure assets screened for climate vulnerability with a flexible and costed response action plan
59	A framework for climate resilient urban and regional planning developed, costed and integrated into the First National Spatial Plan
60	Ksh/year allocated for identifying the vulnerable groups in society who are at risk of climate change impacts
61	% of national climate change indicators for which data have been collected and results reported in appropriate documents at county and national levels
62	Number of ministries at national level providing a budget for climate change adaptation spending (with a breakdown) to the Ministry of Finance and Economic Development (MOFED)
63	Number of ministries at national level that have received training for relevant staff on the costs and benefits of adaptation, including valuation of ecoysystem services

Source: Republic of Kenya (2012a), *National Performance and Benefit Measurement Framework: Section B: Selecting and Monitoring Adaptation Indicators*, Ministry of Environment and Mineral Resources, Kenya.

Sample data sheet for a top-down indicator

TOP-DOWN	Description
Indicator	% of county roads that have been made climate resilient or that are not considered to be vulnerable
Type	Institutional adaptive capacity (top-down)/outcome-based
Level	County
Related action, objective or rationale for measurement	Institutional adaptive capacity (top-down)/process-based/national level indicators numbers 2, 3, 4, 5, 6
Interpretation	Roads (particularly dirt roads) are damaged by heavy downpours and flooding. Culverts and [roads] that are unable to accommodate water flows due to under-specification or poor maintenance can exacerbate flooding. Bridges and embankments may also be damaged, making roads impassable. Roads are vital to the economic and social well-being of the country and damage to them impacts multiple sectors, including agriculture and tourism. This indicator measures the proportion of the road network that is not at risk, either by virtue of its design and location, and hence lack of susceptibility to climate related damage, or because it has been subject to adaptation (vulnerability assessment and improvement) that has increased its resilience.
Unit of measurement	%
Method of calculation	Numerator = length of road that are not at risk + length of road that are at risk but that have been subject to relevant improvements (km) Denominator = total length of road in the county (km)
Frequency of measurement	Annually
Baseline year	2014
Duration of measurement	Long-term
Expected trend w. adaptation	Increase
Target	TBC

Sample data sheet for a top-down indicator *(cont.)*

TOP-DOWN	Description
Responsible ministry/ department/ agency	Kenya Roads Board
Sources of data	Kenya Roads Board
Additional comments	The definition of what constitutes a road (as opposed to a track or other route taken by vehicles) is to be agreed with the Kenya Roads Board.

Source: Republic of Kenya (2012a), *National Performance and Benefit Measurement Framework: Section B: Selecting and Monitoring Adaptation Indicators*, Ministry of Environment and Mineral Resources, Kenya.

Bottom-up vulnerability indicators

During the adaptation planning process, stakeholder consultation identified the need to measure a set of vulnerability indicators to complement the institutional adaptive capacity indicators. The vulnerabilities identified during stakeholder consultation were: rainfall variability and drought, heavy downpours and flooding; sea level rise; and hailstorms and frosts. In total, 62 bottom-up county-level indicators measuring vulnerability were identified. Based on the county level indicators, 27 national level, outcome based indicators were produced, that were subsequently reduced to 10. The objective of these indicators is to measure the effectiveness of local and county initiatives to reduce vulnerability at the national level. Many of the indicators are taken from Kenya's Vision 2030. The Vision 2030 indicators were considered relevant for the NPBMF given the close alignment between adaptation and development. The national and county level indicators are outlined below. A sample data sheet also outlines how the national indicators are measured in practice.

Proposed national level vulnerability indicators

Ref. No.	Description	RVD	HRF	SLR	HF
1	Number of people by gender permanently displaced from their homes due to drought, flood or sea level rise [1, 4, 10, 13, 14, 18, 45, 46, 47]	Y	Y	Y	
2	Number of ha. of productive land lost to soil erosion [4, 6, 7, 12, 17]		Y		
3	% rural households with access to water from a protected source [19, 20, 22]	Y			
4	% urban households with access to piped water [19, 20, 22]	Y			
5	Cubic meters per capita of water storage [18, 19, 20, 22]	Y			
6	% of land area covered by forest [18, 19, 20, 23, 24, 25]	Y	Y		
7	% of classified roads maintained and rehabilitated [33, 34, 35]		Y		
8	Number of urban slums with physical and social infrastructure installed annually [21, 30, 36, 37]	Y	Y		
9	Number of households in need of food aid [1, 4, 10, 13, 14, 18, 45, 46, 47, 54, 55]	Y	Y	Y	
10	Number of County Stakeholder Fora held on climate change [58, 59, 60, 61, 62]	Y	Y	Y	Y

Key: RDV – increase in rainfall variability and drought; HRF – increase in heavy rainfall and floods; SLR – sea level rise; HF – increase in occurrence of abnormally large hailstones/frost in mountain areas.

Note: The numbers in [square brackets] are the reference numbers for county level indicators to which these national level indicators relate.

Source: Republic of Kenya (2012a), *National Performance and Benefit Measurement Framework: Section B: Selecting and Monitoring Adaptation Indicators*, Ministry of Environment and Mineral Resources, Kenya.

Proposed county level vulnerability indicators

Ref. No.	Thematic focus	RVD	HRF	SLR	HF
	Agriculture and rural development				
1	Number of farmers/ fishermen in the county who benefit from credit facilities	Y	Y	Y	
2	Number of farmers/ fishermen in the county unable to access markets/sell produce at a fair price due to climate related effects	Y	Y	Y	
3	Average savings of poor farmers/ fishermen in the county	Y	Y	Y	
4	Average number of suitable crops planted per poor arable farmer in the county	Y			
5	Number of arable farmers in the county on land prone to flooding or landslip		Y	Y	
6	Number of cropping extension workers per farmer working in the county	Y	Y	Y	
7	Number of arable farmers in the county whose land has been stabilised by tree planting, terracing or supporting structures as a result of government intervention		Y	Y	
8	% of tea plantation area that is sensitive to frost damage				Y
9	% of maize planting area that is damaged by hailstones				Y
10	% of poor livestock farmers in the county that keep cattle breeds resilient to rainfall variability and drought	Y			
11	Number of livestock farmers in the county on land prone to flooding		Y	Y	
12	Number of livestock extension workers per farmer working in the county	Y	Y	Y	
13	Ha of alternative (emergency) grazing lands identified for poor livestock farmers in the county on land prone to flooding and drought	Y	Y		
14	% of poor freshwater fishermen on lakes with declining or fluctuating water levels	Y			
15	% of poor sea fishermen dependent on coral reefs that are bleached or at risk of bleaching			Y	
16	Number of fishermen who have been supported by government projects in a switch to sustainable aquaculture/ mariculture			Y	
17	Number of fishing extension workers per fisherman working in the county	Y	Y	Y	
	Environment, water and sanitation				
18	% of arable farmers benefitting from water supplies or irrigation systems designed to alleviate drought problems in the county	Y			
19	Number of water catchments in the county with management plans in place and updated	Y	Y		
20	Volume per capita of portable drinking water in the county	Y	Y	Y	
21	Number of people (by gender) in flood prone areas in the county benefitting from sanitation projects that address flooding		Y		
22	Number of river monitoring stations in the county for which data have been collected	Y	Y		
23	% (by area) of protected areas in the county which have a management plan that addresses climate change	Y	Y	Y	
24	% of people in the county with access to community forest woodland for non-timber forest products (NTFPs)	Y			
25	Ha of gazetted forests, wildlife corridors and dispersal areas in the county	Y			
26	Number of guards per ha of gazetted area for protection/ enforcement of law	Y			
27	Area of coral reef protected from unsustainable exploitation			Y	
28	Number of functional weather stations in the county for which data have been collected				
29	Number of operational early warning systems in the county	Y	Y	Y	Y
30	Average time spent by women collecting water	Y			
	Physical infrastructure				
31	Actual hydropower generated as a % of total hydropower generation capacity in the county	Y			
32	Losses in usable electric power (all modes) due to loss of substations (from flooding) or loss of transmission (from landslides)		Y		
33	Km of county roads that are able to withstand flooding and landslides		Y		
34	Total km of trunk roads in the county that are all weather roads		Y		
35	Number of bridges strengthened or culverts upgraded (or cleared) in the county to cope with higher river flows		Y	Y	
36	Number of people benefitting from flood protection measures in rural and urban areas		Y	Y	
37	Number of households benefitting from slope stabilisation projects in urban areas		Y		
38	Number of people in flood prone areas that receive early warnings of flooding in rural and urban areas		Y	Y	
	Tourism trade and industry				
39	Number of wildlife/ safari tourists visiting the county	Y	Y		
40	Number of beach/ other tourists visiting the country			Y	
41	Investment by county government in measures that protect wildlife (thus enhancing its resilience)	Y			
42	Value of tourist revenues taken in the county	Y	Y		

Proposed county level vulnerability indicators *(cont.)*

Ref. No.	Thematic focus	RVD	HRF	SLR	HF
43	Ksh investment by county government in supporting sustainable tourism initiatives	Y	Y	Y	
44	Number of businesses that have access to risk insurance against extreme weather episodes	Y	Y	Y	Y
45	Number of small, medium and large scale traders whose businesses fail due to climate change impacts in the county	Y	Y	Y	Y
46	Number of ethnic/cultural groups whose livelihoods are lost due to extreme weather conditions	Y	Y	Y	
47	Investment per capita spent by the county government on assisting vulnerable (lost livelihoods) communities	Y		Y	
	Human resource development				
48	Number of people covered by malaria prevention schemes/ treatment facilities in areas that were previously unaffected by malaria		Y	Y	
49	Number of new medical and research facilities that address new emerging diseases as a result of climate change	Y	Y	Y	
50	Number of wards in the county that report health data on a regular (monthly?) basis	Y	Y	Y	
	Research, innovation and technology				
51	Number of primary, secondary and tertiary education institutions in the county that have factored climate change and uncertainty into their curricula	Y	Y	Y	Y
52	Number of education institutions engaged in projects/ programmes that cover adaptation measures	Y	Y	Y	Y
53	Number of people responding to early warning systems		Y		
	Special programmes				
54	Amount of resources (human, technical and financial) mobilised for disaster risk reduction per year	Y	Y		
55	Number of people and livelihoods saved from climate disasters due to rapid response	Y	Y		
56	Average time spent per day by women in productive activities (i.e. income generating)	Y	Y		
57	Number of women farmers (heads of household) in the county who have secure land tenure	Y	Y		
	Cross sectoral				
58	Number of County Stakeholder Fora held	Y	Y	Y	Y
59	Number of local community proposals to address climate related impacts approved by county government	Y	Y	Y	Y
60	% of approved county building designs in which community participation is reflected	Y	Y	Y	Y
61	% of approved county land allocation agreements in which community participation is reflected	Y	Y	Y	Y
62	Number of trained climate change advisers available to county government to mainstream climate change into county planning	Y	Y	Y	Y

Source: Republic of Kenya (2012a), *National Performance and Benefit Measurement Framework: Section B: Selecting and Monitoring Adaptation Indicators*, Ministry of Environment and Mineral Resources, Kenya.

Sample data sheet for a bottom-up indicator

BOTTOM-UP	Description
Indicator	Number of people by gender permanently displaced from their homes due to drought, flood or sea level rise
Type	Vulnerability (bottom-up)/outcome-based
Level	National
Related action, objective or rationale for measurement	Vulnerability (bottom-up)/process-based/county level indicators numbers 1, 4, 10, 13, 14, 18, 45, 46, 47
Interpretation	In severe cases, floods and drought can cause sufficient damage to property or livelihoods to make people permanently homeless. In the case of sea level rise, the most likely cause of displacement is salinisation of soil and/ or ground water and loss of agricultural productivity or water supplies. This indicator addresses gender because the type of response to climate disasters will vary depending on the gender of those affected. In other words, this information will be valuable in planning appropriate action in the future. The indicator requires some disaggregation because it covers both gender and three types of climate disaster. This means that 6 "sub-indicators" will need to be produced, in addition to an aggregate indicator.
Unit of measurement	Number

Sample data sheet for a bottom-up indicator (*cont.*)

BOTTOM-UP	Description
Method of calculation	*Sub-indicator 1:* Number of females permanently displaced from their homes as a result of flood *Sub-indicator 2:* Number of males permanently displaced from their homes as a result of flood *Sub-indicator 3:* Number of females permanently displaced from their homes as a result of drought *Sub-indicator 4:* Number of males permanently displaced from their homes as a result of drought *Sub-indicator 5:* Number of females permanently displaced from their homes as a result of sea-level rise *Sub-indicator 6:* Number of males permanently displaced from their homes as a result of sea-level rise *Aggregate indicato:* Number of people permanently displaced from their homes as a result of flood, drought or sea-level rise
Frequency of measurement	Annually
Baseline year	2013. The indicator can be measured from now because the data for its measurement already exist and do not depend on adaptation actions.
Duration of measurement	Long-term
Expected trend w. adaptation	It is not possible to predict the trend. Without adaptation, the trend is expected to rise. With adaptation, the trend may rise more slowly, or it may fall, depending on the impact of the adaptation measures. It would be useful to use historical data to establish the current trend, which could be used as the baseline.
Target	TBC
Responsible ministry/ department/ agency	Migration and Resettlement Department
Sources of data	Migration and Resettlement Department for data on population displacement and the reason for it KNBS for population data
Additional comments	The same indicator has been proposed as one of the top-down county level indicators, so measurement will be straightforward assuming all counties have done their calculations.

Source: Republic of Kenya (2012a), *National Performance and Benefit Measurement Framework: Section B: Selecting and Monitoring Adaptation Indicators*, Ministry of Environment and Mineral Resources, Kenya.

National Climate Change Adaptation
Emerging Practices in Monitoring and Evaluation

PART II

Chapter 5

Goals and outcomes in Philippines' climate change action plan

The Philippines has developed a National Climate Change Action Plan outlining the country's agenda for adaptation and mitigation for period 2011-2028. The Action Plan identifies seven priority areas: i) food security, ii) water sufficiency, iii) ecological and environmental stability, iv) human security, v) climate-smart industries and services, vi) sustainable energy, and vii) knowledge and capacity development. For each priority area, a results chain has been developed that outlines the ultimate, intermediate and immediate outcomes as well as activities, outputs and complementary indicators. Although the Action Plan includes long-term objectives, it is specified these are not fixed and can be adjusted if the circumstances change.

To ensure that the Action Plan remains relevant, it will be monitored on an annual basis and evaluated every three years. The annual monitoring will help prioritise adaptation needs and inform the allocation of budgets; the periodic evaluations will assess the efficiency, effectiveness and impact of the Plan. These processes will generate valuable information that government officials can draw upon when deciding whether the national approach on adaptation is the right one, if the circumstances that initially informed the Plan have changed, and if adjustments in the plan or the implementation mechanisms are needed.

Ultimate and intermediate outcomes	Immediate outcomes	Output areas	Indicators
STRATEGIC PRIORITY 1: FOOD SECURITY			
Ultimate outcome: Enhanced adaptive capacity of communities and resilience of natural ecosystems to climate change *Intermediate outcome*: Ensured food availability, stability, access, and safety amidst increasing climate change (CC) and disaster risks.	1. Enhanced resilience of agriculture and fisheries production and distribution systems from climate change.	1.1. Enhanced knowledge on the vulnerability of agriculture and fisheries to the impacts of climate change.	Provincial level agriculture and fishery sector vulnerability and risk assessment conducted nationwide. National and provincial agriculture and fisheries climate information and database established. No. of researches conducted on agriculture and fisheries adaptation measures and technologies developed. No. of appropriate CC adaptation technologies identified and implemented.
		1.2. Climate-sensitive agriculture and fisheries policies, plans and program formulated.	Climate change responsive agriculture and fisheries policies, plans and budgets developed and implemented. No. of CC-responsive agriculture-fisheries policies formulated and implemented. Climate change actions – disaster risk reduction performance monitoring indicators developed and implemented. No. and type of risk transfer (e.g., weather-based/index insurance) and social protection mechanisms developed for agriculture and fisheries.
	2. Enhanced resilience of agriculture and fishing communities from climate change.	2.1. Enhanced capacity for CC adaptation and disaster risk reduction (DRR) of government, farming and fishing communities and industry.	No. of farmers and fisherfolk communities trained on adaptation best practices and DRR. No. and type of formal curricula and non-formal training programs developed and implemented for agriculture and fisheries.
		2.2. Enhanced social protection for farming and fishing communities.	No. farming and fishing communities with weather-based insurance. Increase in the no. of small farmers and fisher folk who are credit worthy.

Ultimate and intermediate outcomes	Immediate outcomes	Output areas	Indicators
STRATEGIC PRIORITY 2: WATER SUFFICIENCY			
Ultimate outcome: Enhanced adaptive capacity of communities and resilience of natural ecosystems to climate change *Intermediate outcome*: Water resources sustainably managed and equitable access ensured.	1. Water governance restructured towards a climate and gender-responsive water sector.	1.1. Enabling policy environment for IWRM and CC adaptation created.	Existing water resources management laws reviewed and harmonized. 100% of licensing of water users. Water governance structure streamlined.
		1.2. CC adaptation and vulnerability reduction measures for water resources and infrastructures implemented.	Existing water resources management laws reviewed and harmonized. 100% of licensing of water users. Water governance structure streamlined.
	2. Sustainability of water supply and access to safe and affordable water ensured.	2.1. Water supply and demand management of water systems improved.	No. of site-specific water supply-demand (water balance) studies conducted. No. of water supply infrastructures assessed and climate-proofed. No. of modifications in the processes and demands for water supply systems and users implemented.
		2.2. Water quality of surface and groundwater improved.	Incidence of water-borne CC-sensitive diseases. No. of highly urbanized cities with sewerage infrastructure. No. of household with access to safe water and with sanitary toilets. No. of cities/ municipalities served by sewerage system/septage system.
		2.3. Equitable access of men and women to sustainable water supply improved.	100% water supply coverage of waterless communities. Reduction in climate-related water-borne health risks.
	3. Knowledge and capacity for CC adaptation in the water sector enhanced.	3.1. Knowledge and Capacity for IWRM and water sector adaptation planning enhanced.	No. of staff from key institutions trained as pool of trainers/resources on IWRM and CC adaptation-mitigation. No. of government-academe-CSOs partnerships working on knowledge-sharing. Appropriate technologies on IWRM, CC adaptation and mitigation. Knowledge products produced and accessed by IWRM practitioners at the national and local level. Updated water resources and users database accessible to various users.

Ultimate and intermediate outcomes	Immediate outcomes	Output areas	Indicators
STRATEGIC PRIORITY 3: ECOLOGICAL AND ENVIRONMENTAL STABILITY			
Ultimate outcome: Enhanced adaptive capacity of communities, resilience of natural ecosystems, and sustainability of built environment to climate change. *Intermediate outcome:* Enhanced resilience and stability of natural systems and communities.	1. Ecosystems protected, rehabilitated and ecological services restored.	1.1. CC mitigation and adaptation strategies for key ecosystems developed and implemented.	Hazard, vulnerability and adaptation maps produced for all ecosystems. No. and types of CC mitigation and adaptation measures in key ecosystems implemented.
		1.2. Management and conservation of protected areas and key biodiversity areas improved.	No. and hectares of PA/KBAs protected. No. of ecosystem towns or eco-towns established. Management plans.
		1.3. Environmental laws strictly implemented.	No. of mining operations in protected areas reviewed and temporarily suspended. Solid waste disposal sites in environmentally critical areas (ECA) closed.
		1.4. Capacity for integrated ecosystem-based management approach in protected areas and key biodiversity areas enhanced.	No. of staff in key government agencies trained and implementing integrated ecosystem-based management approaches. No. of Eco-town communities trained on integrated ecosystem-based management. No. of gendered and accessible knowledge products developed and disseminated through various means and audiences (e.g. multi-media, outreach, reports of monitoring, technical reports, policy papers, etc.).
		1.5. Natural resource accounting institutionalized.	Wealth accounts or ENRA integrated in the national income accounts. Policy on ENRA developed and implemented.

Ultimate and intermediate outcomes	Immediate outcomes	Output areas	Indicators
STRATEGIC PRIORITY 4: HUMAN SECURITY			
Ultimate outcome: Enhanced adaptive capacity and resilience of communities and natural ecosystems and sustainability of built environment to climate change *Intermediate outcome*: Reduced risks of women and men to climate change and disasters.	1. Climate change adaptation (CCA)-disaster risk management (DRM) implemented in all sectors at the national and local levels.	1.1. CCA-DRM integrated in local plans.	Vulnerability and risk assessments conducted in all provinces. No. of LGUs with CCA-DRM plans implemented.
		1.2. Knowledge and capacity for CCA-DRM developed and enhanced.	No. of local and community implementing CCA-DRM. No. of CCA-DRM resource networks mobilized. No. of communities reached by IEC program.
	2. Health and social protection delivery systems are responsive to climate change risks.	2.1. Health personnel and communities capacity on CC health adaptation and risk reduction developed.	No. of LGUs with health personnel trained on CC health adaptation and DRR from the provincial down to the barangay level. No. of academic and training institutions with medical and allied health programs integrating CC and DRM in their curricula.
		2.2. Public health surveillance system developed and implemented in all provinces.	No. of community-based public health surveillance system implemented.
		2.3. Health emergency response, preparedness and post-disaster management implemented at the national and local levels.	Health emergency preparedness and response for climate change and disaster risks in place at the national and local levels.
	3. CC-adaptive human settlements and services developed, promoted and adopted.	3.1. Adaptive and secured settlement areas for vulnerable communities and climate-refugees defined.	No. of fisherfolk, farmers, indigenous communities, and informal settler communities in highly CC vulnerable and disaster prone areas resettled. No. of resettlement areas for climate refugees secured from CC-induced conflicts.
		3.2. Population congestion and exposure to CC risks reduced.	No. of LGUs adopting CC-responsive population management to reduce congestion and exposure to CC risks. No. of LGUs implementing a settlement plan.

Ultimate and intermediate outcomes	Immediate outcomes	Output areas	Indicators
STRATEGIC PRIORITY 5: CLIMATE-SMART INDUSTRIES AND SERVICES			
Ultimate outcome: Adaptive capacity of communities, resilience of natural ecosystems, and sustainability of built environment to climate change enhanced. *Intermediate outcome:* Climate-resilient, eco-efficient and environment-friendly industries and services developed, promoted and sustained.	1. Climate-smart industries and services promoted, developed and sustained.	1.1. Enabling environment for the development of climate-smart industries and services created.	Clear national and local policies promoting the climate-smart industries and services formulated and implemented by 2012. Percent increase in the no. of green businesses/enterprises developed and created.
		1.2. Eco-efficient production adopted by industries.	Percent increase in the no. of businesses whose production processes are more environmentally friendly or efficiently using natural resources. No. of companies participating in the SMART Award.
		1.3. IEC and capability building program for climate-smart industries and services developed.	Capacity building program for climate-smart SMEs developed and implemented. Capability building program on GHG emissions inventory and carbon footprint implemented in at least 20% of large and medium industries by 2016. At least 10% increase in the no. of large and medium enterprises adopting climate-smart best practices such as Environmental Management System (EMS), Greenhouse Gas Reduction (G2R), Cleaner Production and Environmental Cost Accounting by 2016.
	2. Sustainable livelihood and jobs created from climate-smart industries and services.	2.1. Increased productive employment and livelihood opportunities in climate-smart industries and services.	Percent increase in the no. of jobs from businesses that produce goods or provide services that benefit the environment or conserve natural resources. Percent increase in the no. of jobs from businesses that involve making their establishment's production processes more environmentally friendly or conserve natural resources. No. of livelihood opportunities and productive employment created from climate-smart industries and services in the rural areas and highly vulnerable communities.
	3. Green cities and municipalities developed, promoted and sustained.	3.1. Infrastructures in cities and municipalities climate-proofed.	No. of critical local infrastructures assessed and retrofitted. No. of local government units implementing CCA-DRM in the issuance of building permits and location clearances.
		3.2. CC adaptive housing and land use development implemented.	No. of cities and municipalities adopting a CC adaptive mixed-use, medium-to-high density, and transit-oriented development. No. of mixed-use, medium-to-high density transit-oriented real estate / community development for urban poor and working families. No. of local governments adopting design for sustainability and green architecture. No. of municipal and city climate-smart sustainability plan developed.
		3.3. Ecological solid waste management implemented towards climate change mitigation and adaptation.	Ecological Solid Waste Management (ESWM) programs established and implemented in all LGUs in accordance with Republic Act 9003 by 2016. Percentage reduction in the volume of and toxicity of wastes disposed. No. of waste disposal facilities located in environmentally-critical areas closed.

Ultimate and intermediate outcomes	Immediate outcomes	Output areas	Indicators
STRATEGIC PRIORITY 6: SUSTAINABLE ENERGY			
Ultimate outcome: Successful transitions toward a climate-smart development. *Intermediate outcome*: Sustainable and renewable energy and ecologically-efficient technologies adopted as major components of sustainable development.	1. Nationwide energy efficiency and conservation program promoted and implemented.	1.1. Government Energy Management Program (GEMP) implemented.	Percentage reduction in government electricity and fuel consumption and expenditure. Percentage reduction in GHG emissions from electricity and fuel consumption in the government sector.
		1.2. Increased private sector and community participation in energy efficiency and conservation.	No. of industries implementing Energy Management Standards under ISO 50001. No. of real estate development adopting green building standards and design for environment concepts. Percentage reduction in energy consumption in the transport, industrial, commercial, and residential sectors.
	2. Sustainable and renewable energy (SRE) development enhanced.	2.1. National renewable energy program and technology roadmap based on RA 9513 and its IRR developed and implemented.	Percentage increase in sustainable renewable generation capacity. No. of sustainable renewable energy development projects implemented. A national sustainable renewable energy program and technology roadmap developed and adopted.
		2.2. Off-grid, decentralized community-based renewable energy system to generate affordable electricity adopted.	Increased percentage of households in off-grid areas using RE systems. Increased no. of off-grid, decentralized RE systems constructed.
	3. Environmentally sustainable transport promoted and adopted.	3.1 Environmentally sustainable transport strategies and fuel conservation measures integrated in development plans.	Percentage increase in fuel efficiency and economy of existing and new vehicles. No. of cities and urban municipalities with formally developed are integrated land use-transport plans. No. of new land developments using integrated mixed-use, medium-to-high density land-use and transport demand management measures. No. of public transport projects achieving transit-oriented development (TOD).
		3.2. Innovative financing mechanisms developed and promoted.	Percentage increase in new investments on EST.
	4. Energy systems and infrastructures climate-proofed, rehabilitated and improved.	4.1. Energy systems and infrastructures climate-proofed.	No. of energy and transport system infrastructures assessed for vulnerability to climate change and disaster risks. No. of CC-risk vulnerable energy and transport system infrastructures retrofitted, rehabilitated and improved.

Ultimate and intermediate outcomes	Immediate outcomes	Output areas	Indicators
STRATEGIC PRIORITY 7: KNOWLEDGE AND CAPACITY DEVELOPMENT			
Ultimate outcome: Enhanced adaptive capacity of communities, resilience of natural ecosystems, and sustainability of built environment to climate change. *Intermediate outcome:* Enhanced knowledge and capacity of women and men to address climate change.	1. Enhanced knowledge on the science of climate change.	1.1. Improved capacity for CC scenario modeling and forecasting.	No. of centers of excellence on CC science established and capacity enhanced. Percentage increase in financing for established centers of excellence.
		1.2. Government capacity for CC adaptation and mitigation planning improved.	No. of vulnerability and risk assessments conducted. No. of gendered capacity building programs implemented. Percentage increase in the no. of trained personnel in key agencies at the national and local level. No. of government agencies complying with GHG emissions reporting requirement.
	2. Capacity for CC adaptation, mitigation and disaster risk reduction at the local and community level enhanced.	2.1. CC resource centers identified and established.	No. of resource centers identified and networked. No. of CC resource networks accessed by LGUs and local communities.
		2.2. Formal and non-formal capacity development program for climate change science, adaptation and mitigation developed.	No. of textbooks for pre-elementary, elementary, high school and alternative learning system with CC concepts integrated. No. of higher education curricula with CC subjects integrated. No. of specialized non-formal training programs on CC adaptation and mitigation developed.
	3. Gendered CC knowledge management established and accessible to all sectors at all levels.	3.1. Gendered CC knowledge management established.	No. of government institutions, centers of excellence and CC resource centers linked to a national web-based CC information hub. No. of gendered and accessible knowledge products for various audience and vulnerable groups developed and disseminated. No. of local institutions and communities accessing gendered knowledge products.

Source: Philippines Climate Change Commission (2011), *National Climate Change Action Plan 2011-2028*, Climate Change Commission, Manila.

National Climate Change Adaptation
Emerging Practices in Monitoring and Evaluation

PART II

Chapter 6

Indicators used to evaluate adaptation in the United Kingdom

The UK Climate Change Act was introduced in 2008. A legally-binding framework on climate change adaption and mitigation, the Act included a call for the implementation of a National Adaptation Programme (NAP) addressing prioritised climate change risks to England. Further, the Act placed a statutory duty on the Adaptation Sub-Committee, of the Committee on Climate Change, to prepare an independent assessment of progress made in implementing the NAP.

The first evaluation of the NAP will be published in 2015. Subsequent evaluations will be published every two years. Since 2012, however, the Adaptation Sub-Committee has been assessing the level of preparedness in responding to some of the priority climate risks and opportunities identified in the 2012 Climate Change Risk Assessment:

- The 2012 assessment examined the risks and opportunities from flooding and water scarcity for households and businesses.
- The 2013 assessment considered what risks climate change may bring to some of the key ecosystem services provided by the land.
- The 2014 assessment focused on the risks to infrastructure, business and public health.

The 2012 and 2013 assessment reports outlined the indicators used to measure changes in climate change exposure and vulnerability as well as the uptake of adaptation actions to reduce impacts. Such indicators, however, were not outlined in the 2014 report. The table below summarises the indicators from the 2012 and 2013 assessments. The arrows indicate the implications of that trend for climate vulnerability. Decision making is also examined to identify incentives and barriers to adaptation.

Indicator type	Indicator of	Indicator name	Trend	Time series
Indicators used to assess risk of flooding (2012)				
Risk[1]	Number of properties (houses and businesses) in areas of flood or coastal erosion risk (not accounting for defences)	Number of properties in river floodplain	⇧	2001, 2008 and 2011
		Number of properties in coastal floodplain	⇧	
		Number of properties in areas at risk of coastal erosion	⇧	
		Number of properties in areas at risk from surface water flooding (1 in 200 year event)	⇧	
	Annual rate of development (houses and businesses) in areas of flood or coastal erosion risk (not accounting for defences)	Rate of development in river floodplain	⇨	2001, 2008 and 2011
		Rate of development in coastal floodplain	⇩	
		Rate of development in areas at risk of coastal erosion	⇩	
		Rate of development in areas at risk from surface water flooding (1 in 200 year event)	⇧	
	Number of properties (houses and businesses) built in floodplain, accounting for defences	Proportion of floodplain development in areas at significant risk of river/coastal flooding	⇨	2001, 2008 and 2011
	Change in hard surfacing	Area of impermeable surface in urban areas	⇧	2001-11
	Vulnerable populations at flood risk	Number of households within highest 20% of ranked deprived communities in areas of significant flood risk (accounting for defences)	⇩	2008-11
		Number of care homes in areas of significant flood risk (accounting for defences)	⇩	
		Number of schools in areas of significant flood risk (accounting for defences)	⇩	
Action	Design of new development in areas at flood risk	Proportion of Environment Agency objections to planning applications on flood risk grounds that are over-ruled by local authority	⇨	2005-10
	Provision of flood defences	Number of households at reduced risk due to construction of new or enhanced defences	⇧	2008-11
		Effective spend in flood risk management activities (capital and revenue) from public and private sources	⇧	2008-11
	Retrofitting property-level measures	Number of existing properties at flood risk retrofitting property-level measures	⇧	2008-11
	Management of surface water in built-up areas	Proportion of new development with sustainable drainage systems	⇧	2008-11
	Provision of early warning systems	Uptake of flood warnings by properties in the floodplain	⇧	2008-11
Impact	Flood damages	Annual insured losses from flooding (UK)	⇨	1990-2011
	Deaths and injuries from flooding	Number of deaths caused by flooding events, per year	⇨	1950-2011
		Number of injuries casued by flooding events, per year	⇨	1950-2011
		Number of mental illness cases caused by flooding events, per year	⇨	1950-2011

Indicators used to assess risk in water scarcity (2012)			Long-term (10yr +)	Most recent year trend (2011 or 2012	
Risk	Supply	Security of Supply Index (SOSI)	⇧	⇧	2002-12
	Overall demand	Freshwater abstraction (non-tidal) by sector	⇨	⇨	1995-2009
	Household demand	Average per capita consumption – all households	⇧	⇩	2000-11
	Household demand	Average per capita consumption – metered households	⇧	⇧	2000-11
	Household demand	Average per capita consumption – unmetered households	⇧	⇧	2000-11
	Agricultural demand	Average volume of water applied for irrigation per hectare by crop type	?	⇧	2005 and 2011
Action	Reducing demand	% of properties with water meters (England and Wales)	⇧	⇧	2000-12
	Increasing supply	Total Leakage (England and Wales)	⇩	⇧	1992-2011
Impact	Water availability (public water supply)	% of reservoir capacity filled (England and Wales)	⇨	?	1988-2009
	Water availability (economic)	Catchments where water is available for abstraction (England and Wales)	?	⇨	2009-11
	Water availability (environmental)	Compliance with Environmental Flow Indicators (England and Wales)	?	⇨	2009-11
	Water availability (social)	Number of drought orders	⇩	⇧	1976-2012
	Water availability (social)	Number of water companies issuing hosepipe bans (England and Wales)	⇩	⇧	1974-2012

Indicator type	Indicator of	Indicator name	Trend	Time series
Indicators used to assess trends in risk and action for key ecosystem services and habitat types (2013)				
Risk	Agriculture – water availability	Total abstraction for agriculture (surface water and groundwater)	⇧	1974-2010
		Total water demand for irrigation	⇩	1990-2010
		Area of crops in climatically suitable locations (potatoes, winter cereals, sugar beet, carrots, spring barley)	⇨	2000 and 2010
		Number of catchments with water available for abstraction	?	2011
Action		Total on-farm reservoir storage capacity	⇧	2007-13; 2005-10
Risk	Agriculture – soil productivity	Total soil carbon concentration in all soils	?	1978-2003; 1978-2007
		Total soil carbon concentration in arable soils	⇩	1978-2007; 1978-2003
		Development of agricultural land	⇧	2000, 2008 and 2011
Action		Uptake of soil conservation measures on wheat and barley fields (only)	⇧	1985-2010
Risk	Agriculture – technological capacity	Total factor productivity of UK agriculture	⇧	1973-2010
Action		R&D spend on agriculture	⇩	1987-2009
		Number of farmers reporting that they are adapting to climate change	?	2011
Risk	Forestry	Percentage of timber trees (oak/beech/pine/spruce) planted in areas likely to be climatically suitable in 2050	⇧	1970-2010
Action		Diversity of species delivered for planting by the Forestry Commission	⇧	2005/06 and 2012/13
Impact		Total forest area impacted by wildfire	⇨	2008-13
Risk	Wildlife	Proportion of Sites of Special Scientific Interest (SSSIs) in favourable condition	⇩	2003-13
		Proportion of SSSIs in unfavourable but recovering condition	⇧	2003-13
		Extent of semi-natural habitats	⇧	1998-2007
		Area of land designated as SSSI and number of protected sites	⇨	2003-13
		Number and condition of "natural connections"	⇨	1998-2007
Action		Area of habitat restoration	⇧	1995-2012
		Area under "landscape-scale" conservation	⇧	1995-2012
Risk	Regulating services provided by upland peats	Proportion of blanket bog SSSIs in favourable condition	⇩	2003-13
		Proportion of blanket bog SSSIs in an unfavourable but recovering condition	⇧	2003-13
		Change in extent of bog habitats	⇨	1998-2007
Action		Uptake of moorland restoration option	⇧	2003-13
		Uptake of catchment-scale restoration	⇧	1995-2012
Risk	Regulating services provided by coastal habitats	Extent of coastal habitats	⇩	1945-2010
		Condition of protected coastal habitats	⇩	1998-2006
Action		Length of coastline realigned (km)	⇧	1991-2010
		Amount of habitat creation, following managed realignment	⇧	1991-2010

1. Indicators of risk includes indicators of exposure and vulnerability

Source: ASC (2013), *Managing the land in a changing climate*, Adaptation Sub-Committee, Committee on Climate Change, London; ASC (2012), *Climate change – is the UK preparing for flooding and water scarcity?* Adaptation Sub-Committee, Committee on Climate Change, London.

National Climate Change Adaptation
Emerging Practices in Monitoring and Evaluation

PART II

Chapter 7

Proposed indicators for monitoring the German adaptation strategy

The German Strategy for Adaptation to Climate Change was adopted in 2008. The objective of the Strategy is to reduce the vulnerability of natural, social and economic systems to climate change and to enhance their ability to effectively adapt to a changing climate. The Strategy includes a risk assessment of 13 action fields and 2 cross-sectional fields that are expected to be positively or negatively affected by climate change. The assessment is complemented by corresponding action points and goals to be developed and implemented together with the Länder and relevant social groups. The complementary Action Plan published in 2011 outlines how the objectives of the Strategy can be achieved. Both the Strategy and the Action Plan are intended to facilitate an integrated approach to adaptation.

An integral component of the Adaptation Strategy is learning through regular assessments of Germany's vulnerability to climate change and the effectiveness of complementary response measures. To achieve this objective, an evaluation framework consisting of three components has been proposed:

- **Vulnerability assessment:** A descriptive evaluation of progress made on adaptation. The assessment will draw on climate projections and information provided by relevant government entities on their awareness of climate change and on their complementary adaptation measures.
- **Indicator-based assessment:** An examination of past and present adaptation initiatives in the 15 action and cross-sectional fields outlined in the Adaptation Strategy. This will be based on an Indicator System approved by the federal government.
- **Evaluation of the Adaptation Strategy:** An evaluation of the extent to which ongoing or planned government initiatives address the projected risks and opportunities from climate change.

The table below outlines the proposed Indicator System.

	Impact indicators		Response indicators
	Action field: Human health		
GE-I-1	Heat exposure	**GE-R-1**	Heat warning system
GE-I-2	Heat wave mortality	**GE-R-2**	Success of heat warming systems
GE-I-3	Contamination with pollen of Common Ragwort	**GE-R-3**	Pollen information service
GE-I-4	Risks from oak processionary moth infestation		
GE-I-5	Vectors of pathogens		
GE-I-6	Vector-born diseases		
GE-I-7	Contamination by cyanobacteria of bathing waters		
	Action field: Building sector		
BAU-I-1	Thermal load in urban environments	**BAU-R-1**	Recreation areas
BAU-I-2	Summer Urban Heat Island (UHI) effect	**BAU-R-2**	Space heating requirements in domestic situations
		BAU-R-3	Funding for climate-adapted construction and refurbishment
	Action field: Water regime, water management, coastal and marine protection		
WW-I-1	Groundwater level	**WW-R-1**	Water exploitation index
WW-I-2	Mean runoff	**WW-R-2**	Structural quality of water bodies
WW-I-3	Flood water runoff	**WW-R-3**	Investment into coastal protection measures
WW-I-4	Low-water		
WW-I-5	Water temperature of lakes		
WW-I-6	Duration of the summer stagnation period		
WW-I-7	Start of the spring algae blooms		
WW-I-8	Sea level rise		
WW-I-9	Intensity of storm waves		
WW-I-10	Seawater temperature		
	Action field: Soil		
BO-I-1	Soil water storage in agricultural soils	**BO-R-1**	Humus contents of agricultural soils
BO-I-2	Rainfall erosivity	**BO-R-2**	Size of grasslands
		BO-R-3	Area of organic soils with natural hydrologic regime
	Action Field: Agriculture		
LW-I-1	Shifts in agrophenological states	**LW-R-1**	Adaptation of management rhythms
LW-I-2	Interannual changes in yield	**LW-R-2**	Cultivation and seed multiplication of warmth-loving crops
LW-I-3	Quality of yield products	**LW-R-3**	Varieties of grain maize categorised in maturity groups
LW-I-4	Hail-storm damages in agriculture	**LW-R-4**	Adapted use of crop varieties
LW-I-5	Pest infestation	**LW-R-5**	Application of pesticides
		LW-R-6	Agricultural irrigation
	Action field: Woodland and forestry		
FW-I-1	Changes in tree species composition in designated Forest Nature Reserves	**FW-R-1**	Area of mixed woodlands
FW-I-2	Endangered spruce stands	**FW-R-2**	Investment into forest conversion
FW-I-3	Incremental growth in timber	**FW-R-3**	Forest conversion of endangered spruce stands
FW-I-4	Infested timber – extent of casual use	**FW-R-4**	Conservation of forest genetic resources
FW-I-5	Extent of timber infested by spruce bark beetle	**FW-R-5**	Humus reserves in woodland soils
FW-I-6	Forest fire hazard and forests / woodlands affected by fire	**FW-R-6**	Forestry-related information on the theme of adaptation
FW-I-7	Forest condition		
	Action field: Fishery		
FI-I-1	Distribution of warmth-adapted marine species		
FI-I-2	Catches of warmth-adapted species in lakes		
	Action field: Energy industry (conversion, transport and supply)		
EW-I-1	Weather-related disruption of electricity supply	**EW-R-1**	Diversification of energy generation
EW-I-2	Weather-related non-availability of electricity supply	**EW-R-2**	Diversification of end energy consumption for heating and cooling
EW-I-3	Coolant-temperature related under-production of electricity by thermal power plant	**EW-R-3**	Facilities for electricity storage

	Impact indicators		Response indicators
EW-I-4	Potential and real yield from wind energy	**EW-R-4**	Water efficiency of thermal power plant
	Action field: Financial services industry		
FiW-I-1	Claims expenditure and claims rate in terms of residential building insurance	**FiW-R-1**	Insurance density regarding extended insurance for natural hazards to residential buildings
FiW-I-2	Loss ratio and combined ratio in residential building insurance		
FiW-I-3	Assessment of one's own insurance cover		
	Action field: Transport, transport infrastructure		
V-I-1	Navigability of inland navigation routes		
V-I-2	Weather-related causes of road traffic accidents		
	Action field: Trade and industry		
I-I-1	Heat related reduction of productive efficiency	**I-R-1**	Water intensity in the processing industry
	Action field: Tourism industry		
TOU-I-1	Bathing water temperature on the coast		
TOU-I-2	Thermal load in spaces used for their healthy climate		
TOU-I-3	Snow cover for winter sports		
TOU-I-4	Preferred holiday destinations		
TOU-I-5	Number of bed nights in coast areas		
TOU-I-6	Number of bed nights in ski resorts		
TOU-I-7	Seasonal bed nights in German tourist areas		
	Cross-sectional field: Spatial, regional and physical development planning		
		RO-R-1	Priority areas and restricted areas reserved for wildlife and landscape
		RO-R-2	Priority areas and restricted areas for the supply of drinking water or use as water reserves
		RO-R-3	Priority areas for precautionary measures against flooding
		RO-R-4	Priority areas for special climate functions
		RO-R-5	Settlement and transport areas
		RO-R-6	Built-over land in areas at risk from flooding
	Cross-sectional field: Civil protection		
BS-I-1	Person hours required for dealing with weather related damaging events	**BS-R-1**	Information on behavior in case of disaster situations
		BS-R-2	Disaster prevention by the population
		BS-R-3	Emergency drills and exercises
		BS-R-4	Persons active in civil protection services

Source: Schönthaler, K., S. Andrian-Werburg and D. Nickel (2011), Entwicklung eines Indikatorensystems für die Deutsche Anpassungsstrategie an den Klimawandel (DAS), Dessau-Roßlau, UBA (updated in March 2014).

National Climate Change Adaptation
Emerging Practices in Monitoring and Evaluation

PART II

Chapter 8

Australia's proposed climate adaptation assessment framework

Australia has proposed a National Adaptation Assessment Framework to assess progress in adapting to the impacts of climate change. The Framework is structured around three sets of questions intended to help shape the response measures needed by business, government and communities:

- What drivers in society and the economy would promote good adaptation?
- What activities would be expected to take place now if Australia is adapting well?
- What outcomes can be expected from good adaptation?

The assessment framework is based on the premise that decisions made today will determine the country's success in adapting to future climate change. It is therefore important that the risks are well understood, and that the governance structure (e.g. building codes, land-use planning and regulation of energy infrastructure) and market mechanisms (e.g. price signals and disclosure of climate risks) facilitate effective adaptation to both climate variability and change. Broad public acceptance is also a pre-requisite for action on climate change adaptation. To assess progress, 12 indicators have been proposed. These are summarised in the table below.

Indicator category	Title	Description
Adaptation drivers	Number of major climate risks satisfying all criteria for good risk allocation	To track progress in understanding climate risks and allocating them to those best placed to manage them, while the beneficiaries from risk management pay the costs.
	Effect of climate hazards on land prices	To measure how climate risk affects land prices for which data is available and climate signals are more likely to be detectable than for other asset, insurance or capital markets.
	Percentage of corporations disclosing climate risk	To track progress in market disclosure of risks from climate change impacts.
	Percentage of the public who accept that some things may need to be done differently in a changing climate	To measure changes in the perception of public and decision makers and their acceptance that it may be necessary to do some things differently in a changing climate.
Adaptation activities	Percentage of organisations considering climate change in long-term planning	To measure if organisations that make decisions with long-lasting consequences (e.g. land use planning, infrastructure) take climate change impacts into account when making decisions with long-term consequences.
	Proportion of tertiary courses in engineering, architecture, planning, natural resource management and other relevant disciplines where climate change is integrated into training	To assess progress in building the skills and information needed to manage risk now and in the future by tracking the extent to which key professionals are being trained to operate effectively in a changing climate.
Adaptation outcomes	Change in the replacement value of built assets in bushfire, flood, coastal erosion and inundation zones	To map the value of assets in climate vulnerable areas and at risk in a more extreme future climate, determining to the value of climate damages (complemented by information about changes in asset design to factor in changes in building design and protective measures in place).
	Damages from natural disasters	To estimate total damages from natural disasters by combining insurance losses from major events with government payments for disaster relief and recovery and estimates of non-insured losses.
	Sensitivity of the value of agricultural production to climate extremes	To measure how much the value of agricultural production declines in response to climate extremes, using a method for comparing the sensitivity of agricultural production to extremes of different severity and areal extent.
	Extent and condition of key climate-sensitive ecosystems	To monitor changes in the condition of key climate sensitive ecosystems (e.g. coral reefs and montane ecosystems) as an indicator of changes in climate risk to natural ecosystems.
Adaptation in the coastal zone	Capacity of planning frameworks to support effective management of climate risks in the coastal zone	To track if coastal planning frameworks take a risk management approach, involve the community, are based on adequate underpinning science, clearly articulate values to be protected in the long term, are developed within a strategic planning framework and provide legal protections for decision-makers acting in good faith based on sound science.
	Number of local governments considering climate change risks in land use planning	To monitor if coastal climate risks – including sea level rise and more intense storm surge – are taken into account in land use planning, development controls and plans for major infrastructure.

Source: Australian Government (2013), *Climate Adaptation Outlook: A Proposed National Adaptation Assessment Framework*, Department of Industry, Innovation, Climate Change, Science, Research and Tertiary Education, Commonwealth of Australia, Canberra.

National Climate Change Adaptation
Emerging Practices in Monitoring and Evaluation

PART II

Chapter 9

Measures and actions in France's national adaptation plan

The French National Adaptation Strategy, adopted in 2006, marked the beginning of the government's focus on adaptation. The Strategy identifies four overarching goals to be considered in national planning processes: i) to protect people and property from the effects of climate change by enhancing safety and public health; ii) to take social considerations into account and to avoid inequality in the exposure to climate risks; iii) to limit the costs linked to the effects of climate change and to exploit possible opportunities; and iv) to preserve French natural heritage.

Complementing the Adaptation Strategy, the first National Adaptation Plan for the period 2011-15 aims to facilitate the planning and implementation of effective adaptation actions and to ensure a coherent approach across areas of public policy. The Plan outlines 19 areas and one cross-sectoral theme considered particularly vulnerable to climate change. For each area and theme, an action sheet outlines a number of actions, each comprising several components that must be undertaken in that area. This totals 84 actions that can broadly be categorised as: i) production and dissemination of information, ii) adjustment of standards and regulations, iii) institutional adaptation, and iv) direct investment. The identified actions (summarised in the table below) facilitate an annual monitoring of progress made in achieving the objectives of the Adaptation Plan.

Action field	Key measure	Action
Cross-cutting	Systematically mainstream climate change in delegated public service contracts and public service contracts let by the government	1. Define climate change reference scenarios 2. Systematically mainstream climate change in delegated public service contracts let by the government 3. Mainstream climate change projections in risk assessments over the life expectancy of classified installations 4. Facilitate thinking in order to define the notion of adaptation 5. Increase research into adaptation in the context of Future Investments
Health	Create a "Health-Climate" monitoring group within the High Commission for Public Health (HC)	1. Consolidate "Health-Climate" research 2. Introduce or increase monitoring of risk factors likely to be influenced by climate hazards (extreme events) 3. Evaluate the risks to human health associated with extreme events and assess the health impacts of adaptation measures, notably by creating a "Health-Climate" monitoring group 4. Develop preventative health actions taking into account the consequences of extreme events and adapt vigilance and alert mechanisms 5. Raise awareness among all stakeholders and provide education via targeted training, information and communication initiatives
Water resources	Develop water-saving and ensure more efficient use of water – make 20% savings in water abstracted, excluding winter water stocks, by 2020	1. Improve understanding of the impacts of climate change on water resources and the impacts of various potential adaptation scenarios 2. Provide effective tools for monitoring structural imbalances phenomena, resource scarcity and drought within the context of climate change 3. Develop water saving and ensure more efficient water use – reduce water abstraction by 20%, excluding winter water stocks, by 2020 4. Support the development of activities and land use which are compatible with locally available water resources 5. Reinforce the integration of climate change issues into water planning and management, in particular in the next water agency intervention programme (2013-18) and programmes for development and water management (2016-21)
Biodiversity	Study the current and potential future consequences of climate change for biodiversity by pursuing and promoting the approaches already initiated in networks of protected areas	1. Integrate biodiversity issues associated with climate change adaptation into research and experimentation 2. Reinforce existing monitoring tools to take into account the effects of climate change on biodiversity 3. Promote integrated land management, mainstreaming the effects of climate change on biodiversity 4. Integrate climate change adaptation into strategies and plans implemented by the government to preserve biodiversity
Natural hazards	Establish an infrastructure designed to acquire, process, archive and distribute sea level data in order to observe and understand long-term sea level variations	1. Develop knowledge (hazards, issues, methods) in the various sensitive areas 2. Extend observation and make data available 3. Standardise the concept of vigilance, alerts and the associated mechanisms and make systematic provision for lessons learned feedback 4. Mainstream the impact of climate change on natural hazards in urban development management 5. Reduce vulnerability and improve resilience and climate change adaptation
Agriculture	Promote water-efficient agriculture	1. Pursue innovation via research and lessons learned and facilitate its transfer to professionals and teachers 2. Promote spatial planning relating to local vulnerabilities and the new opportunities available 3. Adapt monitoring and alert systems to new health risks 4. Manage natural resources sustainably and in an integrated manner to reduce the pressures caused by climate change and prepare for ecosystems adaptation 5. Manage the risks inherent in variability and climate change in agriculture
Forest	Conserve, adapt and diversify forest genetic resources	1. Pursue and increase research and development on adaptation of forests to climate change 2. Collect environmental data, promote it and make it accessible and ensure monitoring of impacts on ecosystems 3. Promote the adaptive capacity of forest stands and prepare the timber sector for climate change 4. Preserve biodiversity and services delivered by forests facing natural hazards 5. Anticipate and manage extreme climate events
Fisheries and aquaculture	Adapt the French shellfish sector to climate change issues	NA

Action field	Key measure	Action
Energy and industry	Promote the use of more efficient cooling equipment (air conditioning) or equipment using renewable or recoverable energy	1. Manage the emergence of peaks in summer energy consumption via an electrical capacity obligation mechanism 2. Promote the use of more efficient cooling equipment (air conditioning) or equipment using renewable or recoverable energy 3. Make all hydrogeological and climate data available 4. Integrate climate change into the monitoring indicators of the Framework Water Directive 5. Identify French industrial sectors which are vulnerable to climate change and potential opportunities (2030-2050)
Infrastructure and transport systems	Review and adapt technical standards for the construction, maintenance and operation of transport networks (infrastructures and equipment) in continental France and overseas territories	1. Review and adapt technical standards for construction, maintenance and operation of transport networks (infrastructures and equipment) in continental France and French overseas territories 2. Study the impact of climate change on transport demand and the consequences for reshaping transport provision 3. Define a harmonised methodology to diagnose the vulnerability of infrastructures and land, sea and airport transport systems 4. Establish a statement of vulnerability for land, sea and air transport networks in continental France and in French overseas territories and prepare appropriate and phased response strategies to local and global climate change issues
Urban planning and the built environment	Reinforce comfortable summer temperature requirements in buildings	1. Incorporate climate change into urban planning documents 2. Adapt nature management and green space management in cities 3. Combat heat waves in cities and reduce the heat island effect 4. Take steps to improve comfortable temperature levels in buildings in the context of a global rise in temperatures
Tourism	Refresh the brand image of cross-country skiing and trekking by mainstreaming sustainable development in ski resorts	1. Promote and develop cycle tourism provision 2. Refresh the brand image of cross-country skiing and trekking by mainstreaming sustainable development in ski resorts
Information	Develop a reference website to disseminate scientific information	1. Increase communications aimed at the general public, elected representatives and business, using as many methods as possible 2. Organise the dissemination of sectoral impacts to prepare the public for adaptation measures 3. Collate and disseminate basic information on climate change, its effects and the adaptation required 4. Raise awareness among decision-makers and provide relevant information to assist them in decision-making
Education and training	Make teaching resources available to the educational community	1. Make teaching resources available to the educational community 2. Gain a more accurate understanding of the impact of adaptation to climate change in each of the areas studied within the framework of the Plan for Careers in the Green Economy and disseminate the results 3. Incorporate health, public health, environmental and occupational health professionals, etc. into the Plan for Careers in the Green Economy in order to provide them with professional training on issues relating to sustainable development in the broad sense of the term and to climate change in particular 4. Provide additional training for business start-up advisors so that climate change is incorporated into analyses of business start-up opportunities 5. Improve ADEME's climate change adaptation external training resources for Regional Climate-Energy Plans (PCET)
Research	Set up an "Adaptation to climate change?" Wiki	1. Improve understanding of climate change and its impacts 2. Support research 3. Develop thematic research projects 4. Promote research
Funding and insurance	Identify and disseminate criteria, methods and data sources so that inappropriate adaptation can be detected	1. Adapt policies, plans, programmes and corporate strategies using sustainable development integration tools 2. Introduce eligibility criteria into the relevant public and private funding mechanisms to avoid inappropriate adaptation projects 3. Mobilise resources for adaptation 4. Provide funding for specialist expertise for small local authorities and SMEs 5. Adapt incentive mechanisms to individuals 6. Improve insurance cover whilst tying it in more effectively to preventive policies 7. Evaluate the costs and benefits of adaptation actions

Action field	Key measure	Action
Coastline	Develop coastal observation networks	1. Adopt a national coastal margin management strategy and develop coastal observation networks 2. Improve understanding of the coastline: the environment, natural phenomena and physical and anthropic development 3. Adapt regulations and forms of governance 4. Reinforce coastal strip management methodology and adapt the various management strategies
Mountain	Integrate a climate change adaptation component into Massif Programmes	1. Mountain agriculture and forests 2. Governance 3. Natural hazards 4. Tourism and leisure
European and international action	Support climate change adaptation in West Africa in the water and agriculture sectors	1. Contribute to developing European adaptation policy and improving regional climate knowledge 2. Increase international cooperation to improve understanding of climate and meteorological and hydrological events 3. Build the capacity of developing countries to prevent the socio-economic risks and impacts linked to climate variability and climate change 4. Provide support for local and regional institutions to promote the integration of adaptation into development planning
Governance	Support the development of regional climate change adaptation strategies	1. Support the development of regional climate change adaptation strategies 2. Support experience sharing in relation to mainstreaming climate change in regional development strategies

Source: French Government (2011), *National Plan Climate Change Adaptation*, Ministry of Ecology, Sustainable Development, Transport and Housing, Paris.

ORGANISATION FOR ECONOMIC CO-OPERATION AND DEVELOPMENT

The OECD is a unique forum where governments work together to address the economic, social and environmental challenges of globalisation. The OECD is also at the forefront of efforts to understand and to help governments respond to new developments and concerns, such as corporate governance, the information economy and the challenges of an ageing population. The Organisation provides a setting where governments can compare policy experiences, seek answers to common problems, identify good practice and work to co-ordinate domestic and international policies.

The OECD member countries are: Australia, Austria, Belgium, Canada, Chile, the Czech Republic, Denmark, Estonia, Finland, France, Germany, Greece, Hungary, Iceland, Ireland, Israel, Italy, Japan, Korea, Luxembourg, Mexico, the Netherlands, New Zealand, Norway, Poland, Portugal, the Slovak Republic, Slovenia, Spain, Sweden, Switzerland, Turkey, the United Kingdom and the United States. The European Commission takes part in the work of the OECD.

OECD Publishing disseminates widely the results of the Organisation's statistics gathering and research on economic, social and environmental issues, as well as the conventions, guidelines and standards agreed by its members.

ÉDITIONS OCDE, 2, rue André-Pascal, 75775 PARIS CEDEX 16
(97 2015 03 1 P) ISBN 978-92-64-22966-2 – 2015-07

www.ingramcontent.com/pod-product-compliance
Lightning Source LLC
LaVergne TN
LVHW081417110826
845149LV00010B/1775
9789264229662